FEATS OF STRENGTH

Feats of Strength

How Evolution Shapes Animal Athletic Abilities

SIMON LAILVAUX

Yale

UNIVERSITY PRESS

NEW HAVEN AND LONDON

Yale University Press books may be purchased in quantity for educational, business,
or promotional use. For information, please e-mail sales.press@yale.edu (US office)
or sales@yaleup.co.uk (UK office).

Set in Electra type by Newgen North America, Austin, Texas.
Printed in the United States of America.

Library of Congress Control Number: 2017959020

ISBN 978-0-300-22259-3 (hardcover : alk. paper)

A catalogue record for this book is available from the British Library.

This paper meets the requirements of ANSI/NISO Z39.48-1992
(Permanence of Paper).

10 9 8 7 6 5 4 3 2 1

For my father, George Lailvaux,
Who was always asking me when I was going to write a book;
I wish you could have read it

CONTENTS

CONTENTS

INTRODUCTION

In 1999, I attended a meeting of the Physiological Society of Southern Africa in the beautiful town of Stellenbosch, South Africa. Although I was there to present the results of my very first research project, a study of the physiology of a little-known species of gecko, I found myself attending a lot of talks on the physiology of sport. The mid- to late 1990s were an exciting time for South African sport. Having recently cast off their international pariah status, South African athletes had burst back onto the world sporting scene, most notably with the victory of the Springbok rugby team at the 1995 Rugby World Cup, and the transition of those athletes from amateurs to full-time professionals was well under way. A key part of that transition was research being done by scientists at the University of Cape Town's Sports Science Institute, and given UCT's proximity to Stellenbosch, many of those scientists were in attendance at that meeting.

During a presentation by one such researcher, I absent-mindedly scribbled a note to myself on a piece of scrap paper, which read, "Athletics and sports science . . . has anyone done this on animals?" and then immediately forgot about it. When I came across the note shortly after my return home, I made my way down to the life sciences library at the University of the Witwatersrand in Johannesburg to investigate. It turned out that athletic-type research *had* been performed on animals, for decades in fact, and it was fascinating. Within a matter of months, I changed my research focus from metabolism to studying locomotor performance in lizards. That turned out to be the first step along the road that led me to my present career and, ultimately, this book.

I consider myself extremely lucky to have a job that allows me the freedom to pursue my research on the things that interest me. For the past seventeen years, those interests have involved, either centrally or peripherally, animal athletic abilities (or *whole-organism performance*, as researchers commonly call it) and have led me to conduct performance research on a variety of animal species spanning four continents. I was also fortunate to have chosen a field of study that has attracted the attention of some of the very best and most exciting thinkers in physiology, ecology, and evolution. The advantage of this is that there already existed, long before I or many of my contemporaries arrived on the scene, a solid understanding of the most basic aspects of animal athletic performance. The physiological and biochemical pathways supporting locomotion; the mechanics and kinematics underlying particular performance capabilities; the energetic costs of flying, swimming, jumping, and running—these things, and more, were worked out in painstaking detail by the pioneers of performance biology, as were many of the mathematical, statistical, and practical techniques for measuring them.

This extensive foundation of knowledge allows us newcomers to apply many of those preexisting tools for understanding animal athletic abilities to a plethora of biological situations involving performance. The result is a research field that is vibrant, ever-expanding, and endlessly enthralling. How do animals use their performance abilities to capture prey and to avoid being eaten? Why does environmental temperature exert such a strong influence on the athletic abilities of certain groups of animals, and how do those animals deal with temperature variation? How does performance change as animals age? Are females attracted to athletic males? How do fish climb waterfalls, snakes glide, and kangaroos travel for free, and what is the airspeed velocity of an unladen swallow (African or European)? These are some of the questions that performance researchers attempt to answer, and the answers have led to a deeper understanding of both the natural world and the process of evolution. Just as often, they have also led to yet more questions.

My desire to write this book grew in part out of the many awkward interactions I have had with nonbiologists who react with almost universal puzzlement and disbelief when they find out what I do for a living. How can chasing a lizard down a racetrack be a job, they wonder? What could you possibly hope to learn from making beetles fight each other? It is apparent that whereas observing the stars, studying volcanos, and smashing subatomic particles into one another are all considered by the general public to be genuine fields of scientific endeavor, measuring how high fleas can jump or how far spiders can run is not. But if the study of animal performance seems frivolous, it lies at the heart of some important concepts and questions in organismal biology.

One of the features of animal athletic performance that attracts so many researchers is its centrality to, and interconnectedness among, several fields in ecology and evolution. Performance is crucial to so many aspects of the day-to-day lives of animals, from reproduction and feeding to signaling, mating, and foraging, that it can be studied from a variety of perspectives. Indeed, performance is so important that it has come to form one of the cornerstones of the evolutionary study of *adaptation*. There is no single, universally accepted definition of adaptation, but a useful way to think of an adaptation is as some feature of an organism that has been shaped by natural selection to aid it in surviving and, ultimately, reproducing.[1] For example, camouflage and cryptic behavior in some species of stick insects are considered adaptations, because excellent camouflage allows individuals to avoid predation and ultimately reproduce more relative to non- or less-camouflaged individuals of the same species. Performance not only meets this adaptive criterion in many types of animals but also comprises several of the most stunning instances of adaptation in the natural world.

Because of the key role that evolution has played in shaping animal athletic abilities, one cannot talk about performance without also considering why such performance abilities have evolved. This book is therefore as

---

1. A more formal definition might be "an alternative among a set of variants that causes maximum reproductive success." Try it out on any evolutionary biologist friends you might have the next time you are in the mood for a bewildering argument.

much about evolution as it is about performance, and I have used performance as a lens throughout to examine several fascinating aspects of the evolutionary process. Consequently, I have organized the following chapters around themes rather than specific performance abilities. Running, jumping, flying, biting, gliding, swimming, climbing, burrowing—they are all in here, and more, but because I want to focus not only on the "how" of performance but also on the "why," they show up as required to illustrate important concepts and ecological and evolutionary contexts.

An important caveat in this regard is that performance is a large and ever-growing field of research. Given the immense difficulty in representing every aspect of modern performance research, I have taken the easy way out and made not even a cursory attempt to do so. Instead, I have allowed myself to be directed by my own eclectic tastes and interests in drawing examples and cases from the literature—at least, for the most part. Some animals necessarily demand more discussion than others. For example, if I appear to talk about lizards a lot throughout this book—and I do—it is not because of any personal bias toward saurian studies or because I just happen to work on lizard performance. Rather, lizards have historically been an important model system for studying whole-organism performance because they are both easy to maintain and measure in the lab and convenient to work on in the field. Consequently, we know a lot more about performance capacities in various contexts for lizards than for any other type of animal. To paraphrase and mangle a far pithier quote on evolution by the great paleontologist George Gaylord Simpson, I don't think lizards are especially interesting because I work on them; rather, I work on lizards because of how much we can learn about performance from doing so. Since I am not the only scientist to think this way, the field of whole-organism performance research is lizard-heavy, and my treatment of the subject here reflects that. (Conversely, there are also many fascinating animals about whose performance capacities we know very little, for reasons that will become apparent. Those animals receive short shrift in this book.)

Yet one further caveat is that I have not been shy about erecting large, flashing signs at several points in the narrative, alerting the reader to things that are currently unknown or poorly understood. Uncertainty sits poorly with those unfamiliar with the scientific process, and there is a common misconception that revisions (or indeed corrections) to previous findings and data mean that science is somehow broken. In reality, the self-correcting nature of science is its greatest strength. That many areas of performance still demand explanation and rigorous experiment does not necessarily mean that my colleagues and I have been slacking or that we are bad at our jobs; rather, it signifies the complexity of nature. From my perspective, the many opportunities for greater understanding of animal performance, though daunting, are tremendously exciting.

Two small notes before we begin. First, a word about units. Measuring performance means putting numbers on it, and the interpretation of those numbers rests on presenting them with the appropriate units. I use metric units throughout this book, mainly because in biology we use metric units exclusively, but also because working with multiple systems of measurement leads to ambiguity. Nonetheless, I present nonmetric units and numbers parenthetically wherever possible. I thus aim to ensure that everyone is on the same page regardless of the preferences some of us may have for various systems of measurement. With regard to the numbers themselves, a quick Google search of animal athletic records brings up a plethora of anecdotal accounts and misinformation. In this book, I draw only from verified, peer-reviewed accounts of animal performance in the published scientific literature, and I highlight the rare occasions where I do not. Although I make no claim to authority on this matter, I do argue for credibility.

Second, the astute reader will have already noted the sporadic appearance of footnotes. These contain information that is technical or tangential, the occasional personal anecdote, and even a few brilliant jokes. Those who don't like footnotes can safely ignore them.

FEATS OF STRENGTH

ONE

Running, Jumping, and Biting

Dawn breaks golden across the Okavango Delta as a female cheetah stalks silently through a fringe of long grass. She moves low to the ground, cautiously, step by careful step. On the savanna beyond, a herd of impala grazes. The herd is restless, and individuals on the edge of the group will occasionally stop feeding, lift their heads, and gaze off into the hazy distance, alert for danger. They do not see the predator shrouded in half-light and plant matter, nor do they smell her; she is all but invisible, her spotted, tawny coat breaking up her outline against the vegetation, and there is no breeze to carry her scent. The sentinels' noses and ears twitch nonetheless—peril is everywhere and can strike at any time. The cheetah's wide, forward-set eyes scan the herd, searching intently. Before too long, she fixes on a target: a subadult impala, not yet properly mature or old enough to know better, which has strayed just far enough from the protection of the herd to have attracted the attention of the hunter in the grass.

The cheetah circles through the grass stalks, powerful limb and back muscle rippling beneath her pelt as she slowly closes the distance on the hapless impala. Finally, she is in position. She crouches down, eyes fixed ahead on her prey, and waits. She remains immobile for almost a full minute before exploding from the grass in a sudden blur of motion. The commotion alerts the sentinels, and all is panic. The subadult is inexperienced but not oblivious; it bolts for the herd without hesitation. The cheetah is closing rapidly, and although the young antelope has already hit top speed, it cannot win on a straight run. The impala swerves to the left, the sudden change of direction intended to throw off the pursuing predator, but the cheetah follows smoothly, swinging her large, broad tail out to the right to counteract her inertia, decelerating and then accelerating again in fractions of a second. The impala jinks again, to the right this time, and again the cheetah follows, flinging her tail behind her to the left. The young antelope's tactics are for naught, and with the swipe of a large, feline paw at the youngster's ankle, the impala tumbles to the ground. The she-cheetah is onto the prone animal immediately, and her jaws clamp around the doomed subadult's throat.

The cheetah's bite is far less powerful than its stride, and it may take up to five minutes for her to suffocate an adult impala in this way. On this occasion, the young animal's suffering is mercifully brief, and it succumbs in just over two. All told, the hunt has taken less than fifteen seconds, during which time the cheetah has covered 173 meters (189 yards) and, on this occasion, achieved a top speed of 25.9 meters per second (m/s) (equivalent to 93 kilometers per hour [kph] or 58 miles per hour [mph]). By comparison, full-grown impalas have a recorded top speed of roughly half that; this subadult never had a chance. The rest of the herd has fled, including a mother who is yet unaware that she has lost her offspring. The female cheetah, however, has secured the survival of her own offspring, at least for the moment, because this prey is not for her but for her four cubs waiting for her in a den not far from the killing field. Still grasping the dead animal's neck with her jaws, she drags it back toward the den. Soon her cubs will feed, and the cycle of life and death will continue.

On the other side of the world, in a public park just beyond the edge of the French Quarter in the heart of New Orleans, Louisiana, a male green anole lizard perches on a broad, green leaf and surveys his territory. This male is large, as green anoles go, and old; a heavyweight, roughly the length of a man's hand from the tip of the lizard's scarred snout to the end of his long tail, he tips the scales at more than 6 grams (g), slightly more than the weight of a quarter. He perches with his belly on the leaf surface and his forelimbs held straight underneath him, lifting his large, arrow-shaped head high off the surface, which gives him an excellent vantage point over the surrounding area. In the vegetation below and around him, a female lays flat against a leaf stem, basking, while another moves slowly through the bush, foraging for prey. They are less conspicuous than the male, and much smaller. The heavyweight moves his head laterally with sudden, jerking motions, and blinks. The loose skin under his throat twitches and expands, unfurling his large, flat, pinkish red throat fan, or dewlap, displaying it prominently for all to see. While the dewlap is fully extended, he bobs his head up and down. He does this several times, following a very specific movement pattern, before fully retracting the dewlap and repositioning himself on the frond, turning to face another direction. He is not displaying at anyone or anything in particular, but he displays nonetheless. This is his territory, and he wants everyone to know it.

He is about to extend his dewlap and bob his head again when something catches his eye. Looking down to another leaf about a meter away, below, and to his right, he sees another male green anole—an invader who has stealthily penetrated the borders of the large male's territory and likely intends to do the same to the heavyweight's females harbored within. The heavyweight male will not allow this; this impudence will not stand. Without hesitation, the heavyweight leaps from his leaf, aiming directly for the intruder, limbs spread and jaws open, exerting a force five times that of his body weight (around 3 newtons [N], or 0.67 pound-force) on his leaf perch as he does so.

The force of his jump, combined with his downward trajectory, allows him to traverse the distance to his rival in less than a second. Although he lands on top of the invader, the recoil of the flexible leaf has caused him to misjudge his jump a hair, and instead of clamping his jaws around the back of the trespasser's neck as he intended, both tumble off the leaf and fall to the ground below, where they land separately.

Neither animal is injured; they are both small enough that such spills have little effect on them. But the intruder lizard now faces a serious problem: an angry, aggressive heavyweight male who is fully prepared to defend his territory and his females. The two males circle each other warily with a peculiar, stiff-legged gait. The heavyweight male has puffed himself up, pumping blood into a sinus running from the top of his head down the length of his spine, thereby erecting a crest along his back that makes him appear larger and more threatening. At the same time, he compresses his torso laterally and flares out his dewlap to add to the effect.

The intruder has done the same, although somewhat half-heartedly; he is eyeing the heavyweight's open jaws, the size of his sporadically extended dewlap, and the large, prominent jaw adductor muscles visible at the back of his mouth. A heavyweight male of this size can generate as much as 15 N (3.37 pound-force) of bite force with those big muscles—more than enough to crush the smaller male's skull. The intruder male has placed himself in real jeopardy; by invading the heavyweight male's territory, he has committed the lizard equivalent of picking a fight with someone who can bench-press twenty-five times his own body weight.

Suddenly, the heavyweight male lunges forward with his jaws, and the smaller male dodges. The heavyweight male attacks again, and the intruder backs up. The decision is made; the smaller male cannot win this fight, nor can he afford to be injured. He turns to flee, and the heavyweight gives chase. The heavyweight mustn't stray too far from his own territory, though: for one thing, he may inadvertently invade that of yet another male and have to deal with the owner; and for another, other males could try to sneak in and have their way with his females while he is gone. Satisfied that the trespasser has been seen off, he gives up the chase, climbs back up into the vegetation, and resumes his vigil.

These two scenarios involve very different animals in disparate parts of the world, yet they share at least one important feature: they both involve individuals relying on their athletic abilities to increase their own reproductive success, or *fitness*. By this, I do not mean their athletic fitness in the "I'm super-healthy and run 5 k every morning before breakfast" sense, but rather *Darwinian fitness* in the evolutionary sense, which refers to the number of offspring that an individual produces over its lifetime.

Evolution is a numbers game, and natural selection favors (selects for) any behavior, ability, or trait that allows individuals to increase their fitness—that is, beget more offspring—relative to other members of the same species. In the case of the cheetah, her blinding speed and superior maneuverability on this occasion enabled a successful hunt, which in turn ensures that her infant cubs will feed and live another day. In doing so, she has preserved, for now, not only their survival but also their chances of growing up to have cubs of their own one day. If they do so, then her fitness is increased relative to that of slower, unsuccessful providers that allow one or more of their cubs to starve. In turn, her cubs would likely also inherit her superior speed and thereby also have high fitness, increasing her evolutionary lineage further.

In the case of the male green anole, he has used his agility and the threat of his superior strength to prevent another male from mating with the females in his territory. Female green anoles produce only one egg at a time, and if any of those eggs are fertilized by some other male, then that is one less offspring that he can count toward his own genetic lineage and one more that counts toward that of a rival male. So by fertilizing as many eggs from as many females as he can, this male is attempting to monopolize a finite resource—female green anole eggs—and increase his fitness at the expense of that of other males.[1]

In both of these cases, individual athletic abilities have made the difference between an evolutionary winner and a loser and are thus favored by natural selection. Interactions like these, played out every day and in a myriad of animal species, shape the evolution of some of the most spectacular athletic abilities on the planet—abilities that biologists collectively call *whole-organism performance*. Animals truly are amazing athletes, and the types of whole-organism performance abilities that researchers study are roughly the same kinds of athletic events that everyone becomes extremely interested in every four years when the Olympics roll around—in other words, dynamic, measurable things that animals do that are generally (but not exclusively) related to locomotion. Animals run, jump, swim, fly, bite, climb, burrow, and so much more, and although few can do all of these activities, many specialize in doing one or more of them spectacularly.

The athletics umbrella for animals is broader than that for humans, and performance researchers often focus on certain characteristics that are not on display at the Olympics. For example, to my knowledge there is no

---

1. Survival is less important to fitness than is reproductive success, and we will deal with this and the unfortunate phrase "survival of the fittest" shortly. For now, suffice it to say that survival is important, but only insofar as one generally has to be alive to be able to reproduce.

official Olympic event that measures how fast competitors use their mouths to suck in fluid, but researchers who work on feeding in certain types of fish measure exactly this. On the flip side, you might also suspect that the Olympic games encompass some events that are too far removed from what animals do for this analogy to work, such as fencing, boxing, or beach volleyball. Although it is true that animals as a rule do not tend to go in for team sports, combat is both important and ubiquitous in the animal kingdom and is not limited to green anoles. Animals may even wield weapons such as claws or horns used specifically during fights with other individuals. Those animal fights are also enabled by underlying, measurable functional capacities such as force production, endurance, or power output and thus are legitimate areas of interest for performance researchers.

The diversity and importance of performance attract the attention of a corresponding variety of biologists. Those in the fields of functional morphology, biomechanics, and evolutionary physiology are interested in the structure-function relations of organisms, the mechanics of performance traits, and the physiological and biochemical pathways that support them. On the other hand, behavioral ecologists and evolutionary biologists have an interest in how performance affects survival, fitness, and evolution. The integrative nature of performance makes it invaluable: performance provides insight into many biological phenomena and behaviors throughout the animal kingdom. Consequently, the versatility of performance research also appeals to integrative biologists such as myself who attempt to fit all of these pieces together with the aim of understanding evolution at multiple levels of biological organization, from an individual's bones and muscles all the way up to populations of animals. My research has explored each of these perspectives at some point or another, although my interests lately are more on the evolutionary side of things.

Machines That Go "Bing"!

If we are to understand animal performance abilities, we must have ways to measure them. For many types of animals, measuring performance is half the fun! When we measure the performance of Olympic athletes, we usually need little more than a stopwatch and a tape measure, or the digital equivalent thereof. When we look to the animal world, though, there are potentially almost as many ways of measuring animal performance as there are types of performance to measure. For example, researchers in the past have measured the sprint speeds of ostriches by chasing individual birds in a Land Rover and matching the car speed (as measured by the speedometer) to that of the animal. A similar technique involving boats was used to measure the flight speeds of butterflies moving over lakes in Central America. These days, biologists try to do things with more precision, and they rely on several tried-and-tested methods that cover the majority of performance traits in nature.

Measuring performance often requires specialized equipment. But because the endeavors of scientists tend to be what is diplomatically referred to as niche, we cannot always order equipment that we need off the shelf and are often forced to custom-build apparatus or, if all else fails, improvise.[2] I have previously used paraphernalia as low-tech as a tub of water, cotton thread, a scale, and two different-sized coffee can lids glued together to measure performance in small beetles. Vendors of certain kinds of specialized scientific equipment do exist, though, and so working on performance often necessitates learning how to operate and use a number of wonderful toys. I will briefly describe some of the most commonly studied types of performance, as well as the most important tools and methods that we use to measure them. Although it is unnecessary to understand exactly how performance is measured to appreciate why it is important, the following nonetheless gives a taste of just a few of the methods that the average performance researcher might use.

Does putting a lizard on a racetrack and measuring how fast it can sprint down that track sound like a fun way to while away an afternoon? Well it is, but it can also grow tiresome when you find yourself doing it for the hundredth time at 3:00 a.m. in a room set to 35°C (95°F)! The traditional

---

2. A colleague tells of a sudden experimental need that forced her to rush into a pharmacy in Italy and demand of a bemused pharmacist in broken Italian: "I need all the Vaseline you have!"

lizard or small animal racetrack uses infrared beams spaced at set intervals along the length of the track to determine the maximum sprint speed of any animal that runs along it. It works on the same principle as certain types of speed-averaging traps on highways: as the animal runs down the track, it breaks the beams; a computer records the times at which successive beams are broken and works out how fast the animal had to have been going to cover each distance interval in the recorded times. Each individual is run three to five times down the track, with time for rest in between runs, and the maximum speed recorded is used for analyses. Racetracks are an easy and efficient means of measuring speed, and I've used several different tracks over the years. The one we currently use in my lab is powered by slot-car racetrack electronics and software that comes with several accompanying race-car sound effects and checkered-flag icons. We can construct similar types of raceways for frogs, fish, and other swimming animals, although in some cases other techniques, such as high-speed cinematography (see below), are preferable.

We can use the racetrack to measure sprint speeds quickly and accurately, but we can't use it to measure endurance. For endurance, we rely on treadmills similar to those that are lined up facing wall-mounted TVs in human gyms. In fact, the treadmill my lab uses for lizard work is a human treadmill that we modified to run at a very slow belt speed, because even though some lizards are exceptional sprinters, most have limited capacities to use oxygen to fuel performance, which means that their *aerobic* (oxygen-supported) endurance abilities tend to be far less impressive. Practically any treadmill can be used for most small to medium animal species, although companies now manufacture treadmills for pets, the existence of which I suspect benefits biologists far more than it does pets. In any case, the type of treadmill is less important than convincing the study animal that running on it is in its best interests, and to this end we often have to build specialized enclosures to keep animals on the belt as opposed to off of it, where any sane organism would probably prefer to be. For my Master's degree research, I put a great deal of time and effort into designing and building a custom lizard treadmill, which my lizards flat-out refused to run on. The lizards I work with now are much more obliging.

Another common treadmill design is the vertically oriented hamster-wheel type, which is useful in particular experimental contexts. However, because the hamster-wheel and flat treadmill setups are different, it is not immediately clear how hamster-wheel endurance relates to treadmill endurance. Although one study suggests that data on maximum aerobic capacity (that is, the maximum rate at which an animal can use oxygen, often the goal of measurement of endurance studies) obtained from treadmills and wheels are comparable for small animals, the different running gaits allowed by flat treadmills versus circular running wheels means that other aspects of endurance could still be affected. Use of these wheels is mostly confined to small rodent-type things like, I suppose, hamsters.

Although many organisms perform perfectly well on a treadmill, not all of them do (fish are notoriously poor runners), and some creative thinking is required to measure flight endurance in birds, bats, and insects or swimming endurance in aquatic and marine animals. For flight, popular methods include wind tunnels and, for small insects especially, tethers, whereby one end of a piece of string is glued onto the back of the animal itself and the other is tied to a vertical post. The insect then flies around the post in a circle until it becomes exhausted, and the time to exhaustion recorded. For estimates of endurance in fish, we can pump water through a chamber at a set rate and use the length of time that the fish is able to keep pace with the current and stay in position as a measure of endurance.

High-Speed Video

If you have ever played with a flip-book or seen how traditional animated hand-drawn cartoons are made, you are already familiar with the concept of frame rate. Frame rate, measured as frames per second (fps), is the number of frames or individual images that a camera records in one second, and altering the frame rate allows us to affect the apparent speed of video playback. In a cartoon or flip-book, static images are displayed in sequence, and doing so at high speeds creates the illusion of movement; the more frames the sequence comprises, the more fluid that movement appears. So it is with video cameras. Although a typical video camera records at either 24 or

30 fps, high-speed cameras can record over a range of frame rates, to an upper limit usually of around 1,000 fps (and some researchers push frame rates up to more than 10,000 fps to record ultrafast animal movements).

The value of high-speed cameras in performance research is that we can film clips of animals doing things that are too fast for the eye to see and slow them way down. If we film at, say, 500 fps, that means that there is a time interval of 0.002 seconds between each recorded frame. Filming at higher frame rates enables even better resolution of movement, at the cost of some clarity in the images (more illumination is required at higher frame rates; alternatively, newer cameras allow the resolution to be maintained at the cost of greatly narrowing the field of view). Playback of high-speed recordings one frame at a time therefore allows us not only to see exactly what the animal is doing but also to measure the position of appendages, centers of mass, or even the entire organism in each successive frame. Through the application of calculus, and with an appropriate reference scale, we are then able to determine from those frame-to-frame changes in position any number of relevant movement variables, such as velocity, acceleration, or mass-specific power output, which we can use to describe individual performance. High-speed cameras are especially useful for species and situations where we cannot measure performance any other way, and they are extremely versatile. Although they can be fiddly to set up and the videos themselves are time consuming to analyze, they reward the patient researcher with lots of high-quality data.

Another feature of modern high-speed cameras is that they can be taken into the field and used to film animals in their natural habitat. Researchers from Cornell University used a high-speed camera to figure out how a group of birds called manakins produce their characteristic nonvocal acoustic signal—a loud snapping sound—by filming displaying birds in nature. They showed that manakins use four distinct signaling mechanisms based on clapping either their wings or their feathers together in various ways over their backs and by flicking their wings rapidly into the air, similar to cracking a bullwhip. It is amazing to think that animals can perform these motions so fast that not only are we unable to discern their movements with the naked eye but we cannot even tell apart four distinct kinds of movements without using high-speed videography! In addition to figuring out baffling bird signals, high-speed cameras are used to measure performance types ranging from sprinting and climbing to flying, swimming, and striking.

Whereas high-speed cameras were once rare and exotic items, these days they are almost commonplace, and high-speed recordings can today be made on GoPros and smartphones. Although high-speed cameras are extremely useful, like any other piece of performance-measuring equipment their successful application in experiment hinges on the cooperation and abilities of the experimental subject. On one occasion when I was a PhD student, my lab mate and I tried to use the lab's high-speed camera to measure the jumping ability of a species of giant grasshopper. Unfortunately, we were unable to get the grasshoppers to really jump, either because they were unusually obstinate and simply refused or because, as we suspected, they were too large and heavy to jump properly. That experiment ultimately produced little more than a handful of super-slow-motion video clips of giant grasshoppers jumping very short distances and landing on their faces, which, while hilarious, did little to advance the sum total of human knowledge.

Force Output

All animal performance abilities involve the generation of forces to power movement, but in some cases it is the force output itself that is of particular interest rather than motion without explicit consideration of the forces involved (also referred to as *kinematics*). Bite force and claw pinching force are two examples of performance abilities where forces are of primary interest, although of course we can also study the kinematics involved. We could estimate force outputs and the functional consequences thereof using high-speed cameras, provided that we know the masses of the animals involved, but it is sometimes more convenient to measure those forces directly.

Performance researchers have an array of often custom-built devices with which to measure these forces. In my lab, for example, we have a force plate, which we use primarily to measure jumping ability in small animals such as lizards and frogs. The force plate uses special crystals (called piezoelectric crystals) that, when deformed, generate electric current proportional to the force applied to detect forces exerted on the plate in the x, y, and z planes of motion. Software then integrates those force measures and uses them, in conjunction with equations of ballistic motion, to calculate other aspects of a jump, such as acceleration, velocity, angle, and distance. Force plates of various types are also used in creative ways and configurations to study locomotion in animals ranging from squirrels to human beings and even elephants and rhinoceroses. Researchers have also used force plates to measure clinging forces exerted by toepad-bearing lizards.

A second useful force-measurement device is the bite-force meter. This consists of two custom-machined metal plates connected to a piezoelectric force transducer (which measures the forces of stretch and compression) in such a way that when an animal bites down on those plates, the transducer is stretched and the force exerted on the plates is displayed on a hand-held charge amplifier. This is probably the piece of equipment that I use most often, and my dissertation work on male combat in Caribbean *Anolis* lizards, which showed that bite force is an important component of male fighting ability in these animals, would have been impossible without it (and lots and lots of duct tape). My collaborators and I have also used these meters to measure the pinching force of the major claw of fiddler crabs and the transducer itself to measure the force required to pull a male fiddler crab out of an artificial burrow, as well as the force required to induce lizards to drop their tails.

More specialized, larger, and more robust bite-force meters exist for larger animals such as bats, but the piezoelectric transducer is also not the only kind that exists; other researchers have constructed bite-force meters out of strain gauges and household materials, and in my lab we use yet another force meter based on thin, flexible, force-sensitive resistor circuits to measure the bite force of small insects such as crickets and grasshoppers that cannot get their mandibles around the bite plates of the regular force meter.

Remote Sensing

The numbers that I presented for the cheetah hunt at the beginning of this chapter are not made up; they are based on 367 runs by three female cheetahs in Botswana, recorded using a combination of GPS (Global Positioning System) technology and inertial measurement units, just like the accelerometers built into most modern smartphones that tell us which way the phone is oriented. In this remarkable study, that equipment was built into collars that researchers then fitted to cheetahs (three females and two males) and used to track their movements and record actual performance data in free-ranging animals in nature, as opposed to in the lab.

Remote-sensing technology represents an exciting frontier for performance work, and in the future we expect to start getting a lot more data that were previously difficult or impossible to obtain on how animals use their performance abilities in various ecological contexts. Within the past several years alone, researchers have used similar setups, often with data beamed to satellites, to understand performance in free-ranging animals from birds to dolphins. I am especially excited about this new era of performance research because, even though fieldwork is fun and everything, nothing is better than having enormous amounts of data collected for you while you are sleeping or at a bar.

Why Study Performance?

So we can put a lizard on a racetrack and measure how fast it can sprint, or film a cricket with a high-speed camera to measure how well it can jump, or use any other number of methods to measure any number of performance types. But why should we care about any of this? On any given day in most major cities, you could wander down to the local racetrack and watch, depending on where you are, horse races, dog races, camel races, or even ostrich races. If you happen to live in Sydney, Australia, you even have the option of watching crab races in any of several bars (or hotels, as Australians call them, probably for the express purpose of confusing foreigners). But although racing crabs in a bar is a perfectly harmless way to fritter away an otherwise uneventful Wednesday evening, why do scientists waste their time studying such trifles?

Research on whole-organism performance has a variety of broad societal benefits, ranging from increased understanding of the natural world to technical advances that affect our lives both directly and indirectly. For example, almost four decades of the study of the mechanics and evolution of animal athletic abilities has led to a variety of innovations that either mimic or otherwise take inspiration from animal functional abilities. New technology such as Geckskin, a super-adhesive and reusable material that works without residue of any kind, was inspired by studies of gecko toepads and clinging abilities—research that grew out of earlier work on the general locomotor performance capacities of lizards. Studies of animal gaits and movement influence the design of robots based on animals as diverse as insects, crabs, bats, and snakes, with potential applications ranging from observation to military and search-and-rescue. These innovations, and many others, can be traced back directly to basic, curiosity-driven science involving running small animals on racetracks, or filming invertebrates fly with high-speed cameras.

Evolutionary biology is in the curious position of being perhaps the only field of scientific endeavor that is forced to defend its very existence on a regular basis, and related animal and plant research is often tarred with the same brush. In 2011, Republican senator Tom Coburn of Oklahoma released a report entitled *The National Science Foundation: Under the Microscope* in which he chided the National Science Foundation (NSF), the federal government agency that funds basic scientific research, for wasting taxpayers' money on research that he personally considered to be silly. One of the studies Coburn highlighted involved performance directly.

According to this report, the NSF awarded more than half a million dollars to Lou Burnett of the College of Charleston and colleagues in 2008 for the sole purpose of making shrimp run on treadmills. Media coverage of Coburn's report zeroed in on this study. Its notoriety grew further thanks first to a YouTube video of scientists in lab coats measuring the endurance capacities of shrimp running on a tiny treadmill and then to a commercial aired by the AARP (American Association of Retired Persons) containing that same video to show retirees what the government was willing to spend money on rather than them.

Both the video and the accompanying soundbite—"shrimp on a treadmill"—were like a gift to critics intent on making these scientists' research look ridiculous. That in itself is not much of an achievement; it is easy to mischaracterize almost any research, regardless of its importance, when you boil it down to a simple phrase stripped of all context. Would anyone have really supported funding to famous ground-breaking endeavors if we had characterized them as, for example, growing mold on a windowsill (which yielded penicillin); figuring out how certain molecules arrange themselves (which allowed us to understand DNA replication); or explaining why we shouldn't feed meat to cows (which led to the discovery of disease-causing prions)?

To be fair, the NSF proposal targeted by Coburn's office did involve putting shrimp on a treadmill, and in this respect the report is not wrong. But this was done, as all components of funded NSF proposals are, for a good reason. The purpose of Burnett and colleagues' proposed research was to examine how changes in the marine environment, such as hypoxic (low-oxygen) conditions, affect the health and ability to fight infections of ocean organisms. Given that shrimp are highly active organisms, examining immune function in shrimp during activity is certainly of interest. As regards hypoxia, the specific type of activity examined should be something that is affected by oxygen levels in the environment—in other words, an oxygen-supported ability such as endurance. These researchers therefore chose to measure shrimp immune function under low-oxygen conditions during endurance-based activity elicited by treadmill running.

In context then, not only is putting shrimp on a treadmill for this project entirely logical, but it is deeply puzzling that Coburn's office chose to highlight this particular study, given that the BP Deepwater Horizon oil spill in April–July 2010, which dumped 4.9 million barrels of oil into already hypoxic parts of the Gulf of Mexico, introduced toxins to a terrific number of marine wildlife, shrimp included, over an area of up to 176,100 km^2 (almost 70,000 sq. mi., or about half the size of Germany). Those individuals along the Gulf Coast who either enjoy eating shrimp or who depend on shrimping for their livelihood—and as a resident of Louisiana, I can assure you that there are a lot of them—might be extremely interested in knowing how long shrimp can run on a treadmill and even more in what this might tell us about their ability to withstand immune challenges and move under low-oxygen conditions. That this point was entirely lost on Coburn very strongly suggests that he either didn't read the original NSF proposal closely or didn't grasp the purpose of the proposed research, and very likely both.

Coburn published an annual "Wastebook" until his retirement from public office in 2015 (although as of this writing Republican senator Jeff Flake of Arizona is continuing the practice). He appears to have had a particular dislike for treadmill-based performance research, singling out another performance study, this time using treadmills to estimate mountain lion endurance, for public scrutiny. Given this continued focus, we might ask exactly how much of a given research grant, in dollar amounts, the performance component specifically of such a study accounts for. In the case of the shrimp study, we can put a dollar value on it. David Scholnick, one of the co-principal investigators on the proposal in question and the man responsible for putting shrimp on the treadmill, stated in an article published by *The Chronicle of Higher Education* that he built the treadmill seen in the notorious YouTube video clip himself, out of spare parts and for a grand total of forty-seven dollars—which he paid for out of his own pocket.

The Toad That Hopped across a Continent

The insights gained by the study of animal athletic abilities are not always immediately obvious. In addition to enabling innovation, work on animal performance has proven invaluable in understanding arguably one of the biggest ecological blunders of the twentieth century: the introduction of the cane toad, *Rhinella marina*, into Australia.

In the 1930s, Australian agriculture faced a growing threat from various pests, including a small insect known as the cane beetle that was devastating the newly imported sugar cane crops throughout the Australian state of Queensland. Given the popularity of sugar cane as a commercial crop throughout many parts of the world, this was not a uniquely Australian problem, and other countries had taken steps to protect their produce, including a relatively new method known as *biocontrol.* Unlike traditional pest-control methods such as pesticides or manual removal that may be expensive, harmful, or otherwise ineffective, the idea behind biocontrol is that populations of pest species can be reduced, if not eliminated, via deliberate introductions of natural enemies to affected areas. The beauty of properly implemented biocontrol is that it is self-sustaining and, ideally, self-terminating, with the biocontrol agent itself following the target species into local extinction once it has depleted its food source.

In the early twentieth century, Puerto Rico, faced with a similar menace to its sugar cane crops, introduced the animal that came to be known as the cane toad to its plantations with the aim of eliminating pests. The scheme appeared to have worked, and the toads were hailed as the ideal organism for eradicating sugar cane vermin. In August 1935, 102 cane toads collected from Hawaii, the site of an earlier Caribbean-inspired introduction, were released in the town of Gordonvale in Queensland. An initial study of the feeding habits of the toads concluded—incorrectly as it turned out—that the toads posed no threat to other native fauna. From 1936 to 1937, thousands more young toads were released into plantations and towns around Queensland.

Over the next eighty years, this mass introduction of the large, nonnative toad had two major effects. First, the toads, having evidently never read the diet study, began to eat pretty much every animal they could find that was smaller than they were (except, ironically, cane beetles, which avoided becoming toad food by moving to the tops of the cane stalks, where the toads could not follow, whereas the immature grubs remained safely underground). Second, the animal's extremely high reproductive rate resulted in a population explosion, and the number of toads grew and spread throughout Queensland and into neighboring states incredibly quickly. From the original introductions into isolated sugar cane plantations in 1935, the toad's range currently encompasses more than 1,000,000 km^2 (621,371 sq. mi.) of both tropical and subtropical Australia, and toads have even reached the town of Broome in Western Australia—3,300 km (more than 2,000 mi.) clear across the other side of the continent. Estimates of the rate of toad dispersal are astonishing. Cane toads began their movement through and out of Queensland at a rate of around 10 km (~6 mi.) per year from the 1940s to the 1960s. A 2006 study estimates the current rate of dispersal at approximately 50 km per year—a fivefold increase in the annual progress of the invasion front.

This incredible increase in cane toad dispersal rate over the past eighty years is driven by their extremely high local population densities. Cane toads create a toxin that they secrete from their skin, and any animal attacking or attempting to feed on them has a nasty day ahead of it. Perhaps in an attempt to make the best of a bad job, Australians have been reported to have tried licking the toads and even drying and smoking the skin to get high off the toxin, but other, nonhuman animals prefer to avoid the poisonous toads. This chemical defense renders the toads all but invulnerable to predators, and with no natural enemies, an extremely catholic diet, and the ability to lay many tens of thousands of eggs per year, they reach extraordinary densities. Estimates place cane toad densities at 1,500–3,000 individuals/km^2 (3,902–7,804 individuals/sq. mi.), for a total Australian cane toad population of at least 1.5 billion individuals. The sheer number of toads was brought home for me in 2008 when I visited the University of Queensland in Brisbane. During a tour of the stately UQ campus given by my host, leading performance researcher Robbie Wilson, I asked him where all of the cane toads I'd heard so much about were. Wilson replied that I should look down. I did so, and as I took a step, a cascade of tiny toadlets jumped away from the spot where my foot hit the ground. These animals' reproductive capacity is such that the lawn was covered with juvenile toads, and although not all of them would survive to adulthood, enough regularly do so to produce the astounding numbers of toads that we see today.

All these creatures living in the same place, however, creates huge incentive for some individuals to move from what must be the toad equivalent of Times Square on New Year's Eve to somewhere else where there is less competition for resources. Once there, the toads reproduce, densities rise, and toads move again. The engine of this dispersal ability is extremely strong positive selection on locomotor performance (which means that high-performance toads have a reproductive, and thus fitness, advantage over poorer performers) over the past eighty years, and studies of cane toad locomotion reveal the signature of this selection on performance. A key predictor of locomotor ability in toads is leg length, as toads with relatively long legs for their body size are better performers than individuals with relatively shorter legs. If the increase in dispersal rate is being driven by selection on locomotor ability, then this signature of selection should manifest as evolution in relative leg length over the same time period.

A series of long-term studies of the invasive toads by Rick Shine's research group at the University of Sydney and their colleagues showed that this is indeed the case. Not only are toads with relatively long legs faster over short distances, but they also move longer distances over time periods from twenty-four hours to three days, strongly suggesting that higher speeds enable faster dispersal. Furthermore, if long-legged, faster toads are being selected for based on their dispersal abilities, then the vanguard of the invasion should comprise more long-legged individuals than further back in the invasion. Again, this is exactly what the researchers found, as the best toad athletes are indeed located at the invasion front. As is their wont, those athletic toads also breed with one another to produce athletic offspring. (Shine's group calls this "the Olympic Village effect.")

These rapid shifts in leg length, locomotor ability, and dispersal rate in invasive cane toads together indicate that selection for locomotor ability drives the evolution of ever-faster dispersal rates in invasive cane toads. Another study examining the skeletons of invasive toads showed that as many as 10 percent of large adult toads suffer from severe spinal arthritis associated with these same factors affecting toad dispersal (that is, frequent movement, long legs, and increased locomotor performance). This means that selection on toad locomotor performance has been so strong that even the risks of developing crooked spines are outweighed by the fitness benefits of increased locomotor-driven dispersal rates. In a third study, researchers showed that endurance is also greatly increased in invasive populations as compared to natural, noninvasive populations, which constitutes further evidence for the notion that locomotor performance is driving the extraordinary toad migration.

In the course of hopping across Australia, the cane toads have left ecological devastation in their path, preying on a variety of unique Australian species such as quolls and antechinuses that are ill-equipped to deal with this leaping amphibian nightmare. They have outcompeted, outbred, and outrun every native frog on the continent, and many other types of animals besides. Modern biocontrol is much more rigorous and careful in its evaluation of potential biocontrol agents, and it is no small thanks to the cane toad that vertebrates are now considered to be off-limits for biocontrol (although some parts of the world are on the verge of relearning this lesson through misguided introductions of mosquitofish as biocontrol agents). Yet the case of the cane toad has also taught researchers another valuable lesson about considering not only the diets but the potential dispersal abilities of prospective biocontrol agents; because as we have just seen, dispersal, like so many other aspects of organismal ecology, is all about performance.

TWO

Eating and Not Being Eaten

Almost every animal on the planet either serves as food for some other animal or derives its food from other animals. Very often, it will do both. Predation is ubiquitous in the animal kingdom, and a diversity of strategies exists by which individuals hunt prey or avoid the hunters. Many of these strategies showcase the inventiveness of evolution, and if you are looking for a particularly striking predation strategy in the animal kingdom, you need not look very far.

My favorite such strategy is the one used by ogre-faced spiders, so called because of their large pair of middle eyes. These spiders are also called net-casting spiders thanks to the unique way that they use their webs to capture prey. Unlike orb-weaving spiders, for example, which weave very large and distinctive webs within which to ensnare any prey unfortunate enough to blunder into them, net-casters use a different prey-capture tactic altogether, one based on nocturnal ambush foraging. The ogre-face constructs a triangular silken scaffold, comprising no more than a few threads of webbing attached to supporting vegetation, from which it hangs upside down like some predatory bungee jumper. But the suspension bridge that it creates with its webbing is not the most remarkable construct that the ogre-faced spider fashions. It also creates another, far more specialized web that it holds out in front of itself with its two forwardmost pairs of legs. This web is a marvel: a tightly woven, yet expandable net that the spider either drops or thrusts on its prey, just as Roman gladiators would to capture opponents in the arena (hence yet another common name for these creatures, gladiator spiders). The net and supporting threads are made from different types of silk, with the net silk comprising hundreds of individual silk filaments that enmesh prey without the aid of any type of glue or adhesive. This gives the net an overall fuzzy appearance, whereas the supporting threads are made from a much thinner, nonfuzzy silk. The spider will hang motionless from its suspension threads, holding the net parallel to the nearest substrate directly below — all the better to get the drop on and constrain prey animals as they pass underneath the lurking predator.

Net-casting spiders base the act of prey capture itself on visual stimuli, and these animals are known to have astonishing visual capabilities, even in almost complete darkness. Indeed, the light-absorption properties of their enlarged middle eyes are up to two thousand times greater in some species of net-casters than that of typical diurnal spiders. In support of this notion of a visual targeting stimulus, there are reports of the spiders placing a single white fecal spot on the substrate immediately below the capture area, the implication being that this spot is used as a reference point for target acquisition — a bull's-eye, if you will. As potential prey items cross the white mark, the spider drops and thrusts its expandable net down on the now-actual prey, pinning it down against the surface and ensnaring the prey in its adhesive web before biting and envenoming it.

The net-caster's extraordinary predation strategy is only one among a plethora of fantastic modes of spider predation. Spiders are superb hunters, and their astonishing catalog of prey-capture strategies is both fascinating and ripe for rigorous functional analysis, including as they do another spider tribe, Mastophoreae, members of which deploy their webs as bolas, hurling them at moths that fly within web-throwing range, just as gauchos do to ensnare much larger prey on the pampas.

Performing for Your Dinner

In the early 1970s a group of American biologists found themselves chasing lizards in the Kalahari Desert. The scientists, among them Ray Huey, Al Bennett, Henry John-Alder, and Ken Nagy, were some of the most prominent pioneers of the modern field of evolutionary physiology, with emphasis on the study of animal performance. Their work, conducted on that and subsequent desert sojourns, resulted in some of the most influential published research in the field of ecology. One of those studies, conducted by Huey and Eric Pianka, considered how animals use their athletic abilities to locate and capture food and divided predation strategies into two categories: ambush foragers and active foragers. Based on observations of animals as diverse as secretary birds, vipers, and lizards, Huey and Pianka's framework applies to a range of animal species.

The first category, ambush foragers, comprises what are called sit-and-wait predators—species that employ a prey-capture strategy based on ambush and/or entrapment. Much like the net-casting spiders, these animals will secrete themselves in an inconspicuous or unexpected location, often facilitated by ingenious and devious camouflage, and wait for victims to present themselves. The actual method of prey capture varies enormously depending on the type and size of prey, the type and size of predator, the environment in which the predator-prey encounter occurs, and any number of other ecological and biological factors. But because surprise is paramount, a common feature of many sit-and-wait predators is that they tend to exhibit exceptional burst-performance capabilities, such as swift strikes or jumps.

Were you to visit the Kalahari Desert yourself and take the time to explore on foot, you might be (un)lucky enough to encounter an exemplar of the sit-and-wait predation strategy. Puff adders are notorious in southern Africa for their remarkably effective camouflage, which enables them to take prey unaware. The success of their ambush strategy is bolstered by their striking performance. Puff adders strike with average velocities of 2.6 m/s (9.4 kph, or 5.8 mph)—among the highest strike speeds recorded for any snake species—and with a very respectable average acceleration of 72 m/s² (161 mph/s). These numbers may not be particularly impressive compared to the striking abilities of some other animals—including humans, with Olympic boxers punching more than three times faster than a puff adder strike—but one must bear in mind the mechanics involved. Specifically, snakes have no limbs, which means that they are both holding themselves up off the ground and accelerating the entire front part of their body forward, including their often-large heads, at a remarkable pace using the snake equivalent of the back, abdominal, and lateral muscles alone. Even though their spine is significantly more flexible than ours, for exactly this reason, this is quite a feat, and if you want to gain an appreciation for exactly how impressive it is, go ahead and try it yourself.

As fast and effective as snake striking performance is, however, it pales beside one of the most incredible performance abilities of any animal on the planet—the strike of the mantis shrimp. Mantis shrimp are distinguished by a number of characteristics, including a visual system that would be the envy even of the net-casting spider. But the mantis shrimp's strike is something altogether its own. In these animals, the second pair of appendages on the thorax has been modified into a pair of raptorial claws similar to those of the praying mantis, hence their common name. Based on the shape and function of these claws, mantis shrimp are divided into two classes: the spearers and the smashers. In spearers, the tip of the claw (or dactyl) is sharp and often spiny, whereas in smashers it is blunt, comparable to the head of a hammer. While both use their claws to strike and incapacitate prey, the strike of smashers such as the peacock mantis shrimp is truly exceptional.

Peacock shrimp specialize on hard-shelled prey, such as snails, crabs, mollusks, and oysters, using their hammerlike claws to strike and break the protective shells to get at the delicious animal within. Shells are engineered by natural selection to withstand all kinds of mechanical abuse, and as such it takes a great deal of force to break apart a shell. This in turn means that the mantis shrimp claw must release energy extremely rapidly to power a sufficiently forceful strike. Peacock shrimp achieve this by means of a complicated latch-and-spring mechanism, enabling them to store elastic energy until maximal contraction of the muscles to power the strike is achieved, just as an archer uses the elastic energy stored by pulling back on a bow to shoot an arrow much faster and farther than any person could throw it. On maximum contraction, the latch is released, and the stored energy is liberated far faster than the duration of muscle contraction.

The net result is that the hammer face of the claw shoots forward at truly astonishing speeds (14–23 m/s, about 50–83 kph or 31–52 mph) and accelerations (up to 10,400 times the acceleration due to gravity), and all in an average duration of just 2.7 milliseconds! For comparison, the lowest speed given here is roughly equivalent to the maximum attained by Usain Bolt during the 100 m sprint, and the highest very nearly corresponds to that which Sammy Hagar claims to be unable to drive slower than. The power of this strike is incredible, estimated at a minimum of 4.7×10^5 watts/kg of muscle (285.5 horsepower/lb.)—hundreds of times higher than that available in even the fastest-known contracting muscles—and is ample to smash apart the shells of prey items.[1] These strikes are so powerful that some larger captive species are reportedly capable of smashing through thick aquarium glass with a single strike.

As amazing as the power of the mantis shrimp strike is, it gets even more so. These shrimp are aquatic, which means that they perform their strikes underwater. The mechanics of rapid movement through water creates yet another phenomenon that spells trouble for any animal on the receiving end of a smasher shrimp strike. The extreme speed of the strike causes the formation of a cavitation bubble between the surface of the claw and the site of impact. Cavitation bubbles form when adjacent areas of fluid move at drastically different speeds, resulting in an area of low pressure between them. Such bubbles are short-lived, and when cavitation bubbles collapse, they release energy very quickly in the form of sound, light, and heat. The shock waves resulting from rapidly collapsing bubbles can damage surfaces nearby; a tiny bubble just 2.7 mm (.10 in.) in diameter collapsing near a wall can generate up to 9 MPa (megapascals) of impact pressure in just five microseconds. This corresponds to 1,305 pounds per square inch (psi), roughly the atmospheric pressure on the surface of Venus, which over that short time frame is sufficient to damage boat propellers.

The bubble formed by the face of the mantis shrimp claw striking an animal shell also collapses, and the subsequent shock wave can be as forceful as the strike of the claw alone—sometimes even more so. To put exact numbers on it, peak claw impact forces in the peacock shrimp are in the range of 400–1,501 N (90–337 lb. force), up to more than 2,500 times their body weight. Peak cavitation forces average only half of this, but the very highest are within the range of the impact forces. In fact, cavitation forces are so

---

1. I've fudged the ecology here; technically, spearer shrimp are ambush foragers, whereas smasher shrimp are active foragers. However, the mechanics of both strikes are similar, and from the perspective of a mollusk probably any kind of attack is an ambush.

great that some shrimp hunt using just cavitation alone: snapping shrimp stun prey by shooting cavitation bubbles at them. Peacock shrimp, however, place their prey in the position of double jeopardy, delivering two extremely forceful blows separated on average by only 390–480 microseconds. If we now take into account that each peacock shrimp wields two of these smashing claws, a hapless snail or limpet could find itself the target of no fewer than four shell-shattering wallops, and all within fractions of a second.

Just as mantis shrimp have harnessed the material properties of water in their foraging strategies, so other animals have evolved to use the characteristics of the aquatic medium for prey-capture purposes. Any fish that embarks on an ambush-foraging lifestyle, for example, will face some of the same immediate challenges as snakes—primarily, a notable lack of limbs. Unlike most snakes, however, fish exist almost exclusively in aquatic environments, and although I am not aware of any species of fish that make use of cavitation bubbles for hunting, a popular predation strategy for fish involves the opposite use of water, that is, pulling it toward them as opposed to pushing it away.

Breathing in most fish entails drawing water into the mouth and expelling it through openings in the side of the head covered by bony plates called opercula. (Some fish, such as sharks and tuna, instead use what is called the ramjet principle to force water over their gills as they propel themselves forward.) As the water is being expelled, it moves across the surface of the gills, and here is where the gaseous exchange of oxygen (from the water to the gills, and thus the bloodstream of the fish) and carbon dioxide (from gills to water) occurs. The need for fish to suck water into their mouths turns out to be handy for predatory species, and most bony fish use suction in combination with their striking abilities to capture elusive prey items.

Possibly the most singular suction feeders are found in the Syngnathinae, a subfamily of fish comprising seahorses, seadragons, pipehorses, and pipefish. Among the most beautiful and bizarre members of the animal kingdom, seahorses and pipehorses are often found anchored by their prehensile tails, unique among fishes, to the same aquatic plants that species such as leafy seadragons mimic so spectacularly. These locations are strategically chosen not only to conceal them from passing prey but also to give them a foundation from which to launch strikes at prey items. Prey capture in seahorses, called pivot feeding, is a two-stage process: the animal first rotates its head upward, bringing its usually bent head more in line with its trunk, which has the ultimate effect of bringing the elongated snout closer to the prey item. Once the snout is near enough, the seahorse sucks the prey rapidly into its mouth.

Analysis of the feeding motions shows that the characteristic S-shape of seahorses which gives them their common name is likely an adaptation for pivot feeding, because it allows the animal to swiftly straighten its body and strike at prey from further away compared to pipefish that exhibit a similar behavior but lack the horse-shaped body plan. Similar to the mantis shrimp, syngnathid fishes overcome the inherent power constraints of muscle by storing and releasing elastic energy in the neck tendons and trunk muscles, allowing them to snap their head forward and toward prey items at far higher speeds than if the strike were powered by muscle alone. Studies of juvenile seahorses also show that these animals perform rapid pivot feeding within the first day of birth (technically, expulsion from the brood pouch of the male parent, who carries the eggs) and that the speeds of strikes in young seahorses rival those of striking mantis shrimp—a truly remarkable feat.

The elongated snout of seahorses is also considered to be an adaptation to pivot feeding, but it's not a patch on the elaborate jaw system of some species of bony fish. Suction feeding in some species, such as tropical *Epibulus insidiator*, involves a phenomenon innocuously termed "jaw protrusion." A better clue as to what this is can be gleaned from *E. insidiator*'s common name: the slingjaw wrasse. The jaw anatomy of the slingjaw wrasse comprises a specialized four-bar linkage system that allows these fish to do something truly remarkable that has to be seen to be believed: they extend their jaws outward toward a prey item and then suck the prey into their mouths once the jaws are nearby. To be clear, I am not referring to a system akin to that of xenomorphs in the *Alien* franchise, whereby the mouth opens to reveal a second mouth that is fired toward its target; rather, *the actual jaws shoot forward from the cichlid's face,* supported by the four-bar system that extends like a trellis, and then folds back up again as the jaws are retracted to lie back flush with the snout! (Incidentally, moray eels do have a second set of jaws inside their throat that extend forward to grasp prey items already inside the mouth and transport them back toward the esophagus. Furthermore, the designers of *Alien* likely drew inspiration from a truly disturbing crustacean parasite of fish that eats its victim's tongue *and then replaces it*, attaching itself to the ruined stump by its hind legs and acting as a surrogate tongue replete with its own pair of jaws. So that's out there.)

Jaw protrusion is practiced by a variety of ray-finned fish, but none have taken it to the extreme of the slingjaw wrasse. How many fish must have ended their existence thinking that they were out of range of that larger predator lurking over in the seaweed, only to find themselves being sucked into oblivion as it shot its freakish jaws at them? Peter Wainwright of the University of California, Davis, whose research group works on suction feeding, has several high-speed videos on his website of the slingjaw wrasse and other fish feeding in this manner. I highly recommend them for encouraging undergraduates to stay awake during lectures and for reminding them that evolution routinely devises scenarios more outlandish than anything Hollywood could possibly imagine.

Before moving on, there is one more animal I want to mention that exhibits an especially spectacular mode of ambush foraging. The eel catfish, a denizen of muddy swamps in central Africa, hunts its aquatic prey using a method of suction feeding similar to that of syngnathids and the slingjaw wrasse. But unlike those organisms, the eel catfish refuses to be limited by the cuisine on offer in the rivers and streams it inhabits. A large proportion of this creature's diet is made up of insects, and terrestrial insects at that. That's right: this aquatic fish propels itself out of the water, grabs prey items off dry land, and recedes back into the stygian depths from whence it came to digest at its leisure. What is more, the eel catfish is not the only fish species known to exhibit terrestrial feeding; similar behaviors are thought to have evolved at least five times independently within ray-finned fishes alone.

This raises several questions as to exactly how animals that are adapted to aquatic suction feeding can achieve such a feat. The same tricks that work in the water won't work on land; air is approximately eight hundred times less dense than water, and frictional forces experienced by prey due to air flowing over them are about fifty times lower than they would experience in water, which means that eel catfish cannot possibly be sucking terrestrial insect prey into their mouths from a distance without ramping up their sucking abilities to an impossible degree. These animals therefore have to do something entirely different. They essentially have evolved two separate suites of biomechanical feeding adaptations: one for the aquatic environment and the other for terrestrial situations.

The challenges of terrestrial feeding for the eel catfish stem from its morphology, or shape. Propelling itself out of the water and achieving the requisite speeds to pull off a terrestrial ambush is greatly aided by the sudden transition from dense water to considerably less dense air, and the performance boost thus achieved is based on the same principles that allow flying fish to jump out of the water and gain sufficient altitude to glide on elongated fins (about which more in Chapter 6). However, once out of the water in the head-up position, the eel catfish then has to bend its head downward to grab its prey—something of a problem for an animal that doesn't have a neck. High-speed camera footage collected by Sam Van Wassenbergh and colleagues at the University of Antwerp in Belgium shows that the catfish overcome this problem by bending the head and body laterally, a behavior that is not seen during aquatic feeding. However, the decreased resistance of air relative to water does allow the eel catfish to open (and presumably close) its mouth 50 percent faster when capturing terrestrial prey. Its preexisting suction abilities are not wasted, however, and these same analyses also show that although the catfish may not be able to use its suction abilities to pull terrestrial prey items toward it as it does in the water, it nonetheless may employ suction to draw the captured prey further into the mouth, in the same way that we might suck in strands of spaghetti. Yet another terrestrial-feeding fish, the fascinating mudskipper, will venture onto land with a mouth full of water. It then uses its suction abilities to manipulate prey items inside its water-filled mouth, a phenomenon that researchers have likened to having a hydrodynamic tongue!

The second general category of predation strategies identified by Huey and Pianka is the active foragers, which traverse a particular area searching for prey items that they may even capture while on the move. These animals are far more dynamic than the sit-and-wait predators, and they seldom remain in the same spot for long during a foraging bout. If ambush foragers get by on deceit, surprise, and remarkable burst performance abilities, active foragers rely more on endurance abilities. Indeed, actively foraging lizards that show higher movement rates and activity levels also tend to have enhanced endurance abilities, as well as locomotor gaits that are more suited to slower speeds.

Increased movement rates come at a cost of increased energetic investment in locomotion to fuel that active lifestyle; however, these expenditures are compensated for by higher prey-capture rates, at least in part because active foragers occasionally encounter areas where lots of potential prey items can be found in the same place. Indeed, predator foraging mode appears to be complemented by movement rates of prey, in that active foragers happen upon clumps of sedentary prey that do not move very often, such as termites, whereas sit-and-wait predators that remain in the same place encounter prey that are themselves active and exhibit high movement rates.

Although active foraging animals are not necessarily using burst abilities to leap out from the shadows at prey items, not all active foragers move slowly as a rule. For some types of animals, the performance capacities of active foragers are still spectacular. Raptors are among the fastest animals on the planet and adopt locomotor-based predation strategies that are anything but pedestrian. However, merely being fast isn't often enough, and understanding how animals use their performance abilities to capture prey is just as important. Scientists have begun to attach cameras onto aerial predators, which has broad societal benefits not only in adding to our knowledge of raptor ecology but also in producing videos that are perfectly suited for inducing motion sickness in the viewer. This research shows that falcons, for instance, approach their potential prey from a constant angle, keeping it in view while remaining out of sight, all the while converging not on the spot where the prey is but on where it soon will be. The result is a very efficient predation strategy, whereby the raptor heads off its prey based on the prey's current trajectory rather than wasting energy following it or spiraling in on it, just as one might head off a zigzagging toddler on the playground. From the perspective of endurance rather than pure speed, African wild dogs are able to run down much faster prey because they can run for longer at speeds that the prey cannot sustain. So although an escaping antelope, for example, may have the upper hand at first, if the hunt continues for long enough, the antelope will eventually slow from fatigue and be captured by the loping pack of predators.

Prey-capture efficiency is important because active foraging can be energetically expensive. If active foragers are constantly on the move, and if they are capturing and eating prey over a lengthy time as they go, then the mass of that captured prey ultimately adds to and increases the weight of the foraging animal itself. Because it costs more energy to move a heavier animal than a lighter one over a given distance, an important question that arises when considering active foraging is, "How does the mass of consumed prey affect performance abilities?"

Wolf spiders are a family of spiders that don't build webs and instead rely on their locomotor abilities to run down and capture prey. In these animals, burst speed performance sets an upper limit to how much individuals consume while foraging, and spiders refuse large prey items because consuming them would necessarily lower their sprinting speeds. This phenomenon appears counterintuitive; after all, if the point of foraging is to find food, what is the sense in turning it down once you have it in your grasp? In the case of the wolf spiders, experimental work by Jonathan Pruitt of the University of California, Santa Barbara, shows clear evidence of positive selection on burst speed, meaning that faster animals are more likely to survive than slower ones. The loss of performance as a result of consuming large prey items is a powerful incentive for curbing one's appetite, and so even though it is true that having food is better than not having food, the consequences to wolf spiders of gluttony are dire indeed.

A follow-up question might be, "How large a prey item is too large?" The threshold for prey size beyond which consumption decreases performance enough to become maladaptive probably varies among animal species depending on their size and means of locomotion. Swimming, for example, is a much cheaper means of locomotion than running on land, and feeding experiments using an actively foraging fish, the crucian carp, show that feeding up to 4 percent of body mass results in a decrease in stamina of

12 percent. However, this species also allocates a remarkably high amount of its daily energy budget toward digestion, which means that it processes its food extremely quickly, possibly as a form of compensation. As a result, this performance decrement, such as it is, occurs over a short duration and probably has little effect on the carp's ecology and survival.

Last, it is worth noting that although the two-category predation classification proposed by Huey and Pianka is valuable in a heuristic sense, researchers have suggested many intermediate categories over the years. The dichotomous foraging mode paradigm relies heavily on ecological work carried out on lizards, yet some lizards defy easy categorization as either sit-and-wait predators or active foragers. Indeed, Huey and Pianka themselves explicitly considered flexibility in foraging mode from the start. For example, chameleons might be expected to be the quintessential sit-and-wait predators, given both their penchant for camouflage and crypsis and their famed method of capturing unaware prey by shooting their tongues at them. However, work on foraging behavior in the chameleon *Bradypodion pumilum* by Marguerite Butler of the University of Hawaii at Manoa found that this species exhibits movement rates more consistent with those of active foragers. Furthermore, some species of tuna switch between active and ambush foraging strategies depending on the distribution and availability of prey.

Cases such as these are a cogent reminder that classification systems based on behavior, and sometimes even something as fundamental as species identity, are constructs that exist more for our convenience than as reflections of the natural world. Nature steadfastly resists pigeonholing, and evolution is always capable of producing something surprising.

Keep Yourself Alive

The endgame for any organism is to reproduce, and thus natural selection acts very strongly on animals' reproductive capacities. However, although evolution really is more about reproduction than about survival, it is entirely possible for selection to act on survival, too, because in the same way that dead men tell no tales, dead animals produce no offspring (unless they have managed to inseminate a female shortly before becoming dead). Keeping oneself alive is therefore definitely in an organism's best evolutionary interests, and an important goal of performance researchers is understanding how natural selection acts in those scenarios. It's a tough life as an intermediate-to-lower link in the food chain, and constant peril has proven an effective spur to the evolution of adaptations to steer clear of the clutches of sneaky ambush predators and speedy active foragers alike.

The simplest performance-based antipredator strategy is to run away, and many animal species certainly subscribe to this philosophy. For some species, though, especially those on the smaller end of the mouse-to-elephant size continuum, not only may they not be fast enough to escape a predator in this way, they also may never be. Smaller animals attempting locomotor escape from a predator come face-to-face with an inconvenient fact of biomechanics: smaller animals are slower than larger animals (in absolute terms) almost every time. What to do, then, when faced with galloping, imminent death in the form of a larger, faster predator?

Small, jumpy-type animals such as jerboas and kangaroo rats have evolved a solution based on unpredictability. They use their expert jumping capacities, ably supported by a suite of musculoskeletal adaptations suited for exactly this purpose, such as stiff tendons attached to large, leverlike feet, to jump around at high accelerations in more or less random directions, the idea being that if they do not know which way they are going to go next, the pursuing predator is not likely to either. Thus, any predator attempting to predict an escaping jerboa's future location based on its current trajectory is likely to be blindsided by a sudden and rapid change of direction. Because sudden directional changes are harder to achieve for larger animals (they are more at the mercy of inertia than smaller animals), random changes in trajectory are an effective escape tactic.

Talia Moore based her doctoral work at Harvard University on jerboa locomotion, and her videos of field assistants attempting to catch speedy, unpredictable little jerboas in nature clearly demonstrate the efficacy of this strategy (and call to mind the Benny Hill theme "Yakety Sax" when I have seen them). To test whether jerboa jumps really are unpredictable, Moore applied information theory (which deals with the encoding and

33

transmission of information in patterns and sequences) and the physical concept of entropy to her analysis of jump patterns exhibited by jerboas when escaping predators. Her analyses confirmed that such jump patterns have high metric entropy—that is, the information contained in them is essentially random, and thus a jerboa's future position cannot be anticipated from its current or past locations.

Other potential prey organisms use similar unpredictability-based tactics to avoid becoming actual prey. For example, flying insects that find themselves in the figurative cross hairs of a bat often escape by altering their flight patterns in various ways depending on their distance to the predator. Nocturnal insectivorous bats hunt using echolocation and determine the spatial position and distance to potential prey items by producing high-frequency sounds that reflect off of flying insects, thus betraying the prey's position.[2] In the roughly fifty million years since bats have taken to the night skies, insects have evolved a battery of countermeasures, ranging from specialized ears that warn them when they are being targeted by bat sonar to the ability to produce high-frequency sounds of their own that jam bat echolocation signals by swamping out the echo of the bat's call (hawkmoths do this using their genitals!). Simple neural circuitry in crickets initiates correspondingly simple evasive flight behavior when bat sonar is detected, causing them to steer away from the sound source. When exposed to intense sounds in the frequency range commonly occupied by bat sonar, bush crickets immediately dive; however, the direction of the dive bears no relation to the direction from which the sound emanates, again suggesting that the crickets are choosing a movement direction at random, just as jerboas do.

Perhaps the most sophisticated antibat predation countermeasures are found in moths. Based on the nature of the bat call, moths can determine how far away those bats are and adjust their escape accordingly. Indeed,

2. Insectivorous bats appear to operate under the reasonable assumption that anything small and airborne at night is a flying insect. I once exploited this assumption for my own amusement on an uneventful evening at a research station in Puerto Rico by sitting on the patio throwing peanuts into the air and watching bats swoop down to capture them. I later learned that shaking keys influences moth flight behavior, as it reproduces some of the sound frequencies that bats use for echolocation.

some noctuid moths can detect bats over as much as ten times the distance at which bats can detect moths. Moths far from the source of the bat call simply turn away, whereas those close by zigzag, perform loops, or power dive. Below a certain threshold distance, some moths exhibit a behavior euphemistically called flight cessation; that is, they stop flying, fold their wings, and drop out of the sky. Green lacewings—which are not moths but also exhibit this behavior—take it a step further. If a bat approaches a falling lacewing, as evidenced by a sudden increase in the detected bat call rate, the lacewing suddenly flips its wings open, halting its plummet for just a moment, then folds its wings again and resumes its dive. Researchers sometimes refer to the *evitability* (as opposed to the inevitability) of bat avoidance behaviors, a reference to their unpredictability—again, a feature that serves nocturnal flying insects well in terms of staying out of the bellies of nocturnal insectivorous bats.

While behaving randomly is an effective escape strategy for some species, others use another approach straight out of *Monty Python's Flying Circus*, which famously demonstrated the value of not being seen. Predator avoidance tactics based on camouflage and stealth (grouped under the umbrella term of *crypsis*) are popular throughout the animal kingdom. Crypsis may be forced onto individuals by necessity. Pregnant females, for example, often adopt cryptic tactics because bearing the additional mass of eggs or fetuses makes rapid escape from imminent predation attempts difficult. Crypsis can also be driven by environmental factors, and cryptic species are often those that live in complex environments, such as leaf litter, or that mimic a specific type of microhabitat, such as some anoles that mimic the twigs they live on.

A further general trend that I don't believe has received serious empirical attention is that animals committed to the cryptic lifestyle seem to be characterized by less-than-impressive performance abilities (sit-and-wait ambush foragers notwithstanding). Such a trend, if it exists, might be predicted on the basis of efficiency. The physiological machinery that supports animal performance is expensive both to run and to maintain, and why would you maintain a high-performance engine for which you have no use? Only supercar owners know for sure.

Catch Me If You Can

The logic of crypsis is compelling. Even species that rely on locomotion for escape and that are superlatively adapted to swiftly vacating the immediate premises would be better off if they didn't have to run away. Escaping from predators still incurs costs, ranging from the energetic costs of rapid escape to the opportunistic costs of lost foraging or mating prospects. Imagine just having found a patch of edible grass before anyone else has, or putting a lot of effort into convincing a female that you are the male she should be copulating with right now, only to have to put all of that on hold to run away from some interfering, charging predator. Even if you do survive, there is no guarantee that that grass patch or female will still be available once the predator has moved on. In cases like this, it is worth persuading the predator in question that trying to capture you is more trouble than it's worth. For this reason, certain animals have evolved signals (usually auditory or visual ones) that biologists suspect are intended to communicate to predators the futility of their foraging attempts.

A famous example of signaling to predators is the behavior known as pronking (if you are South African) or stotting (if you are not).[3] Both refer to the same remarkable gazelle display that is widely considered to be a pursuit-deterrent signal directed at other animals intent on killing and eating them (fig. 2.1). Following the sighting of a predator, animals such as springboks and Thomson's gazelles perform a particular jump, with their legs held straight beneath them, their backs arched, and their heads angled downward, almost as if they are practicing some bizarre kind of aerial yoga. Gazelles do not stot in place either, and they will often stot while escaping or being chased. Despite the unusual kinematics, stotting gazelles can attain impressive amounts of altitude (anecdotal estimates report them jumping as high as 3 m [9.8 ft.] and as far as 14 m [46 ft.], although it should be stressed that these measures are unverified), and this is almost certainly

---

3. "Stotting" comes from a Scottish word meaning "to walk jauntily." "Pronk" is an Afrikaans verb meaning "prance" or "strut."

Fig. 2.1. A springbok stotting. Photo by iStock.com/johan63.

by design because this display, after all, is meant to be seen—and it is the predators that are meant to see it.

There are no fewer than eleven hypotheses attempting to explain the purpose of this display, but the one that currently has the most empirical support is that stotting is a signal of unprofitability. Specifically, by demonstrating their good condition and physical (as opposed to Darwinian) fitness, gazelles are communicating to predators that it is not worth making the attempt to catch them, for at least two reasons: first, because the stotting gazelles are particularly healthy and quick and therefore likely to outrun them, and second, because the predators have been sighted and the gazelles are now onto them. Support for this explanation is in the form of data showing that wild dogs preferentially attack Thomson's gazelles that either do not stot or stot less than other individuals. Interestingly, the type of predator seems to matter: gazelles are less likely to stot at stalking predators, such as cheetahs and lions, which get as close as possible to their prey before attacking, than they are at coursing predators, such as wild dogs, which rely on endurance to run down and outlast escaping prey, often beginning pursuit in full view of that prey. This suggests that gazelles are specifically advertising their endurance abilities, because there is little point in advertising this to cheetahs, which rely on speed and tend to hunt over short distances.

If stotting is a signal of unprofitability for the actively foraging predator, then there should be a link between stotting and individual athletic ability. A direct test of this hypothesis would involve measuring the sprint speed and/or stamina of individual gazelles as well as their stotting rates and testing whether one predicts the other. Such a study has yet to be conducted, for understandable logistic reasons, but it is now more feasible than ever thanks to advances in GPS technology that obviate the need to annoy Tom Coburn further by bringing gazelles into the lab and running them on treadmills. Existing evidence of the relation between stotting and performance is indirect at best, because Thomson's gazelles have been shown to stot less in dry seasons, when food is scarce, compared with the wet season, when resources are more plentiful, the implication being that the animals are likely to have less muscle for supporting performance when resources are scarce and that they therefore stot less. However, without hard data this is mere conjecture, and there could be many other reasons. But if linking displays directed at predators to athletic ability has proven difficult in gazelles, this link has been established in a very different animal—a lizard.

In the forests of Puerto Rico can be found, without too much effort, a lizard named *Anolis cristatellus*. Like other anoles it is a beautiful and fascinating animal, and also like most other anoles, it possesses a large repertoire of displays. A mainstay of almost all anole displays is the push-up. This is exactly what it sounds like: anoles will bend their legs and then straighten them, literally doing push-ups just as you would be made to do after talking back to your drill sergeant. Anole displays incorporate push-ups in multiple ecological contexts: males display them to other males during aggressive interactions, to females during courtship, and in undirected territorial displays to no one in particular. They also have been reported to display at potential predators. Although every anole biologist has been the target of an anoline display at one time or other, this was considered by most to be a case of mistaken identity on the part of the lizards. It wasn't until Manuel Leal (then a graduate student at Washington University in St. Louis) decided to test the idea that these lizards were treating humans as potential predators, and thus trying to convey the same message to us that they are signaling to other predators, that we gained insight into what that message might be.

Leal quantified the types of displays produced by each individual lizard in the same predator-prey context by staging interactions between A. *cristatellus* males and a model of a snake predator. He then linked the frequency of push-ups to a particular performance ability called *distance capacity* (also known as maximal exertion). Distance capacity or maximal exertion is measured by chasing a lizard around and around a circular racetrack until the creature becomes very tired and falls over. In other words, it measures how far and how long an animal can sprint rapidly until it can't run anymore—certainly, a relevant metric in an animal that deals with predators by running away from them. Leal's data showed convincingly that lizards that performed more push-ups during the staged interactions also had higher exertion capacities than lizards that performed fewer push-ups. This strongly suggests that A. *cristatellus* males communicate information on their physical fitness, and thus their ability to run away for longer, to predators via push-up displays, much as stotting is thought to do.

Displays toward predators are not limited to gazelles and lizards. And they are also not necessarily visual. Macropodid mammals, for example, a group of marsupials that includes kangaroos, wallabies, and the wonderfully named rat-kangaroos (not to be confused with kangaroo rats), produce an audible alarm call when exposed to predators by striking the ground with one or both feet, like some antipodean version of Thumper. As for stotting, nine competing hypotheses currently exist to explain the purpose of foot-thumping in macropodids, although unlike stotting, the notion of foot-thumping as a signal of unprofitability has received little attention. Similarly, calls produced by birds and other auditory-signaling animals are often linked to certain aspects of individual physiology and function, but links to specific athletic conditions are seldom considered. It is entirely possible that such alarm calls directed at predators are indeed signals of athletic ability, just as they appear to be in A. *cristatellus* and probably are in gazelles, but we currently lack the data to evaluate this idea.

Run for Your Life

Some animals, either fortuitously or by design, adopt the ultimate antipredator strategy: they live in places where predators do not. But what happens if an ecologically naive species suddenly meets a novel, perhaps introduced, predator? Evolutionary theory, intuition, and evidence all show that such a species has but two options: adapt or die.

Selection can act strongly on performance, especially in situations that are literally life or death, and certain species that exhibit neither high performance capabilities nor camouflage in the absence of predators had better evolve one or the other very quickly if they want to stay alive. But although we know quite a lot about how performance capacities enable animals to stay alive in contemporary situations, understanding the conditions that led to the evolution of particular performance capacities is another issue. Evolutionary ecology is in many ways a historical science insofar as we must often try to understand traits and scenarios that are the endpoint of extremely long-term processes of selection and change, the origins of which are not always immediately apparent. However, in some cases it is possible to gain insight into how these processes get their start through the use of cunning experiment.

A fascinating experimental example of predator-driven performance adaptation in action comes from a study of a species of lizard on several small islands in the Bahamas. Islands have a long and storied history in evolutionary biology, even going so far as to be known as evolution's laboratories. This is because islands are discrete areas of sometimes simple species assemblages and communities, the characteristics of which can be understood or manipulated with relative ease. Another useful feature of islands in general, and archipelagos such as the Bahamas in particular, is that experiments conducted on one island can be replicated on another. Replication gives us some idea of whether the outcomes of manipulations that we apply to islands are repeatable or due only to chance, and thus of how much confidence we should place in those outcomes.

The specific Bahamian islands in question harbor yet another species of *Anolis* lizard by the name of *Anolis sagrei*, also known as the brown anole on the grounds that it is an anole that is brown. *Anolis* is well known in biological circles because it is among the most species-rich of the vertebrate genera. There are almost four hundred described anole species to date, many of which occur in the Caribbean and are extremely well studied, only in part because herpetologists were quick to realize that the Caribbean is an awfully nice place to do fieldwork. *Anolis* lizards in the Caribbean have an idyllic existence. Throughout the Greater Antilles in particular they are the only game in town as far as lizards go, and with few exceptions they have a monopoly on the daytime lizard niche in the Caribbean. They also have to deal with relatively few predators. This opportunity likely contributes to the enormous diversity of anoles throughout the Caribbean, and these lizards have radiated and spread into a variety of habitat types, from the tops of trees to their twigs and trunks and to bushes. The brown anole prefers to oc-cupy low trunks and vegetation and is often found doing its lizard thing on the ground itself. This is especially the case when there are no other lizard species around for them to concern themselves with, and whereas large islands may be lousy with various anole species, small islands frequently house brown anoles, and only brown anoles.

Into this lizard paradise a team of scientists led by Jonathan Losos and Tom Schoener introduced not a snake but carnivorous, ground-dwelling curly-tailed lizards. With the permission of the Bahamian government, these insatiable little beasts were released deliberately onto several small Bahama islands populated by brown anoles as part of an experiment aimed at understanding the anoles' evolutionary responses to a novel predator. As a lizard predator, curly-tails are a good choice; not only are they found nearby and regularly colonize neighboring islands, they are also voracious and ag-gressive creatures apt to eat anything they can get their mouths around, and as such they were likely to give the smaller brown anoles a run for their money—which they did.

After releasing the curly-tails, which they presumably did while rubbing their hands together and cackling maniacally, the researchers left the lizards to work out their differences by themselves. Six months later, they returned to see how the brown anoles had fared against the curly-tail onslaught and to measure the survivors. They found that the surviving lizards that had escaped the jaws of the curly-tails had longer limbs, on average, compared to the average limb length in that same brown anole population pre-curly-tail blitz, indicating strong and positive natural selection on brown anole hindlimb lengths. This makes sense if we think about the curly-tail introduction from the brown anole's point of view: you are a small brown lizard, wandering around your tiny island doing small brown lizard things, when one day a bunch of ravenous terrestrial predators appears out of the blue, all of them intent on making you lunch. Since they are bigger and stronger than you, your best option in this scenario is to run for your life. Because long-legged brown anoles are faster than shorter-legged individuals (as is also the case in most other lizard species), only brown anoles with long legs could run fast enough to avoid becoming a snack for the marauding curly-tails. It was a clean and intuitive result, and if Losos and colleagues had stopped there they would have had a simple case study of a population changing in response to a novel predation pressure. Fortunately, the lizard conspirators didn't stop there, and what they found on their next Bahamian trip six months later was even more interesting.

Following their initial sojourn documenting selection on limb length in brown anoles, the researchers returned to the island of traumatized lizards to see whether selection had continued to act in the same way on the anoles to keep them ahead of the curly-tails. However, when they measured the brown anoles again six months later, they found an entirely opposite result to the first measure: selection had reversed direction and was now favoring anole lizards with short limbs over long-legged individuals! This sudden reversal in the direction of selection might seem very odd indeed if we didn't know much about anole ecology. Fortunately, when it comes to anoles, we know quite a lot, and the reason for this change in selection pressure becomes apparent once we consider what the animals were doing. The switch in selection was precipitated by a change in behavior. Running away, it seems, isn't the best way to escape rampaging curly-tails. Instead, the brown anoles had realized almost immediately that the safest course of action was to leave the ground entirely. As a result, over time they had abandoned their initial strategy of running away a lot and begun to take refuge far more often in the trees.

Brown anoles, like most other anoles, have toepads, which in combination with their claws makes climbing much easier for them compared with lizards that lack toepads—such as most other nongecko lizards, including curly-tails. In fact, curly-tailed lizards can only climb very broad trees, and not very well at that. So the brown anoles went up into the trees where the curly-tails could not follow. But crawling around on narrow branches in trees presents very different functional challenges and therefore requires different, shorter limbs to sprinting on open ground. Once the brown anoles had made it safely into the trees, they faced a new problem—their long limbs, which carried them swiftly away from the gaping maws of the curly-tailed lizards, were unsuitable for maneuvering on narrow substrates like twigs and branches. Consequently, now that they had adopted this new, mostly arboreal lifestyle, selection was favoring brown anoles with short legs! The lesson here is that selection can fluctuate rapidly, far faster than anyone had imagined. Had the anole research team waited twelve months to remeasure the population instead of six, they would have missed this sudden and stunning reversal.

Jaw-Dropping Ballistic Ants

To finish off this chapter on eating and not being eaten, I want to consider a special animal that exhibits a particularly interesting athletic ability. Arguably the most astonishing predator strike, antipredatory defense, and extreme performance ability on the planet are all embodied within a single trait: the jaws of the trap-jaw ant, *Odontomachus bauri*.

As entomologists, ant enthusiasts, and both fans of the fourth Indiana Jones movie are aware, ants are among the most rapacious predators in the animal kingdom. Of all the ants, the trap-jaw exhibits the most spectacular predatory strike. This creature holds its enlarged mandibles spread apart very much like an open bear trap, from which it derives its name, and it is capable of firing them at incredible speeds. The descriptors of trap-jaw strikes boggle the mind: the mandibles have been recorded to snap closed with an average speed of 38.4 m/s (138.2 kph, or 85.9 mph), with peak closing speeds ranging from 35.5 m/s (127.8 kph, or 79.4 mph) to an incredible

64.3 m/s (231.48 kph, or 143.8 mph)—just short of the maximum speed attained by the fastest rollercoaster in the world. In a study of jaw function in trap-jaws, Sheila Patek of Duke University (who is also responsible for the mantis shrimp work mentioned earlier) and collaborators also measured the accelerations involved in such extreme performance, and they arrived at an estimate of approximately one hundred thousand times the acceleration due to gravity—among the most extreme acceleration capacity of any animal on the planet. Let us liken this again to human Olympic boxers, who are capable of throwing a punch with a maximum velocity of 9.14 m/s (32.9 kph or 20.4 mph) and a maximum acceleration of only fifty-eight times the force of gravity. Compared to the strike of the trap-jaw ant, the fastest punch of an Olympic boxer may as well be moving in slow motion. Even the mantis shrimp are left in the dust, with these ants clocking speeds approaching three times higher than smasher claw strikes. Unlike an Olympic boxer, however, and like both peacock shrimp claws and seahorses, the mandibles of the trap-jaw achieve such incredible velocities and accelerations not through pure muscle power alone but thanks to a latch-and-spring mechanism. This is a common and recurring theme in performance: whenever you find an animal doing something at gob-smackingly high accelerations and velocities, an elastic storage mechanism is sure to be at work (I will have more to say about this phenomenon in Chapter 7).

Whether powered by muscle alone or enhanced by elastic storage, the strike of trap-jaw ants is amazing enough that these creatures would be forever enshrined in the annals of top animal performance if prey capture were all that the strike was used for. Remarkably, however, these ants also use their superlative mandible strikes for another purpose—ballistic locomotion. I mentioned that trap-jaw ant strikes are also an antipredator defense, and they indeed use exactly the same strike mechanism to escape dangerous situations. To do this, they direct their strikes either straight down at the ground or directly at the predator, launching themselves considerable distances into the air and away from the immediate source of danger.

Patek's research group has performed detailed kinematic analyses of trap-jaw ballistic jumping and has found that there are two distinct antipredatory behaviors involving the mandible strike. The first, called the bouncer defense, involves striking at the predator, which achieves two things: first, the predator is struck with the fastest-closing mandibles on the planet, which has to surprise it at the very least; and second, thanks to Newton's third law, which states that for every force there is another that is equal in magnitude but opposite in direction, the ants are propelled backward away from danger at high speed (1.7 m/s; 6.12 kph, or 3.8 mph) and acceleration (680 times the force of gravity) to an average distance of 22 cm (8.7 inches) away . . . an impressive distance for a tiny insect.

The second defense, called simply the escape jump, is brought about by the ant positioning its head straight down at the ground with its mandibles spread and cocked and then releasing the jaw latch mechanism. This causes the animal to shoot up into the air at an average angle of 76 degrees and up to a maximum height of 7.3 cm (2.87 in.). If you are going to watch just one high-speed video of an ant using its jaws to escape, it should be one of O. bauri. The effect is even more incredible when multiple animals perform this behavior at the same time, resulting in, to quote Patek, "ants bouncing through the air like popcorn." Think about this the next time you trip over a slightly elevated patch of ground; a trap-jaw ant could have cleared that, and it could have done it with its jaws.

Lovers and Fighters

Humans are peculiar animals. Relative to the majority of other animal species on earth, a lot of the things we do that we consider to be normal are in fact quite odd. Many of those things relate specifically to sex. For example, it has been noted many times that humans are among only a handful of species that have sex for pleasure as well as procreation. In fact, we enjoy sex so much that we frequently go to great pains to circumvent the whole procreation part of sex so that we can get on with getting it on without worrying about blowing the household budget, contributing to overpopulation, or having to shop for little shoes. This deliberate avoidance of the fitness benefits to sex is something that most other animals would shake their heads at in bewilderment, provided they belong to a species that has proper heads and also clearly identifiable necks. For other species, the act of sex is a means to an end, and successful reproduction is so critical to fitness that they will risk inconvenience, discomfort, and injury to achieve it. For members of certain species, the males are also willing to fight, kill, and even die for it.

Picky Females and Pushy Males

Males and females find mating partners in nature via a process called *sexual selection*. Unlike natural selection, which governs anything that contributes to lifetime reproductive success (including, as we have seen, survival), sexual selection is concerned with mating and reproduction to the exclusion of all else.

Natural selection favors anything that increases an individual's fitness relative to other members of the same species (be they alligators or blue cranes or elephants), whereas sexual selection favors those traits that result in higher individual fitness relative to other members of the *same sex within that species*. For example, while natural selection will select for the best blue crane (in terms of fitness) relative to all other blue cranes, sexual selection will favor the fittest male blue crane relative to other male blue cranes. As such, natural selection often selects for traits that help animals to feed or escape predators, both of which have the net effect of keeping those animals alive at least until they can reproduce. But because competition between males or between females to obtain mates is so intense, sexual selection selects for traits that increase immediate individual reproductive success even at the expense of survival and longevity. This means that although natural and sexual selection share the ultimate goal of increasing lifetime reproductive success, the two processes do so in different ways and can be entirely at odds.

The notion of sexual selection, which was proposed by Charles Darwin in 1872, explains the existence of male characteristics such as bright colors, long tails, or loud calls that are entirely useless for survival, if not obviously harmful. These types of *secondary sexual traits*, so called because they are usually distinct from the primary sexual traits, such as genitalia, arise in response to one (or, less usually, both) of the two major processes of sexual selection.

The first of these, called *female choice*, selects for the evolution of traits that females are most likely to pay attention to and be drawn to, such as calls of a specific frequency or feathers of a particular color. Females evaluate those traits and base their decision whether to mate with the bearer on the characteristics thereof. The second drives the evolution of weapons or signals that males use in fights with rivals to isolate one or more females from the amorous intentions of other males via *male combat*. Some of these secondary sexual traits, especially those that are subject to female choice, affect the performance of individuals that have them. The classic example is the widowbird, males of which have extremely long tail feathers because females prefer male widowbirds with long tails over those with shorter tails. The female preference for long tails is so strong, and the reproductive benefits of long tails are so large, that male tails can grow long enough that they impair flight performance, causing problems in all of the non-female-choice, survival-related areas. But performance is arguably more directly relevant to the second sexual selection area of male combat, where males use their athletic abilities to fight other males over access to females or resources that females require.

Male combat is ubiquitous within the animal kingdom, but it doesn't always look the way we might expect. When humans think of combat, we think of overly muscled individuals beating each other into submission using either one or more body parts or a weapon specifically designed for that purpose. Although animals do sometimes fight in this way, winner-take-all punch-ups are the exception rather than the norm. The observation that fighting is a dangerous business for both combatants is neither novel nor insightful, just as it is so obvious as to be barely worth noting that being injured sucks. But these truisms have important implications for how male combat happens in nature. If combat were to escalate to no-holds-barred death matches every time, then a given male might indeed win a fight against a rival. However, animals in nature cannot visit the emergency room to be patched up like we can, and the injuries the victor would almost certainly suffer on the way to victory could easily curtail his own survival or fitness even if he was somehow able to avoid future conflicts. There is, after all, no point in winning a fight over a female if it leaves you so wounded that you can't mate with her after you've won.

Full-on, intense violence is in no one's best interests. For this reason, the males of many animal species have evolved ritualized combat, often involving displays and signals, that males use in specific ways to intimidate each other without necessarily progressing to physical combat. Fights between males therefore frequently resemble bizarre performance art or interpretive dance much more than they do the nontalking parts of kung fu movies, and can be likened to the prefight posturing between the gaudily dressed Apollo Creed and stone-faced Ivan Drago in that classic of American cinema *Rocky IV*, or even to the traditional Maori war dance (or haka) performed by the New Zealand All Blacks before their international rugby matches.

How males display differs tremendously among animal species and need not necessarily be visual, with animals using auditory, chemical, and even tactile modalities to bluster and harangue each other. For instance, jumping spiders drum their legs against dry leaves, producing a rapid, staccato sound that is detected by other males, whereas male water beetles defending territories vibrate their bodies, producing ripples of particular frequencies that other males sense.

Bite Me

For these displays to be meaningful, however, they do need to be backed up by physical ability to inflict injury. A population where individuals posture harmlessly at each other and disputes never escalate to physical altercations is one ripe for exploitation by individuals willing and able to use violence as a means of mediation.[1] As such, displays tend to be staged in a stereotyped behavioral sequence that begins with ritualized display and culminates in escalated physical combat buttressed by individual performance capacities.

Combat in lizards, for example, commonly follows a set of rules collectively described as a *sequential assessment game*. This means that fights start off with displays that are easy to produce and that convey some information regarding the individuals and their physical capabilities. In collared lizards, these displays involve the males opening their mouths and displaying to their opponents the large jaw adductor muscles within that power biting. These muscles are substantial, and this, coupled with their overall large heads, means that collared lizard jaws are dangerous weapons. This display is the lizard equivalent of strutting down the beach flexing and kicking sand at another male, and it is both a threat and a warning. If the other male is not intimidated, however—perhaps because he himself is a large male who is not to be trifled with and has large jaw muscles of his own—then combat may advance to the next stage.

This next stage consists of a lateral display, whereby the males compress themselves, as if they are being placed on their sides within the middle of a book that is then closed. This compression is coupled with a specific positioning whereby they face their opponent side-on. The net effect is to make

---

1. The philosophical complexities of this scenario were thoughtfully explored using explosions in the Sylvester Stallone/Wesley Snipes documentary *Demolition Man*.

themselves look larger than they are and ultimately to convince the opponent to back down and disengage. If he does not, then combat escalates to the third and final stage of physical confrontation. In this escalated state of all-out combat, the lizards grab each other by their jaws and mouth-wrestle, or each may attempt to seize the other at the base of the neck. This can continue for some time, until one male eventually gives up and retreats. In several lizard species, males that can bite hard for their size are more likely to win a fight against a relatively weaker biter. In collared lizards specifically, work by Jerry Husak of the University of St. Thomas in Minneapolis and collaborators showed that males that bite relatively hard have more offspring than weaker biters. Hard-biting collared lizards are thus able to access and mate with more females than wimpy biters, resulting in positive selection for bite force in males. Although this relation could also come about if bite force contributed to survival via foraging, Husak and others have also shown that male bite force does not affect survival. Furthermore, while female bite forces are sufficient to crush all of the prey that males eat, male bite forces are much higher than they need to be for foraging purposes. The collective evidence therefore points to male bite forces, enabled by large adductor muscles, having been selected specifically for male combat in collared lizards (and most likely in many other lizard species as well).

Although the ability to bite hard may make you a badass among lizards, escalated combat is rare and tends to occur only in individuals that are very evenly matched for displays and, especially, body size. But such fights, when they do occur, are not benign and can result in serious injuries for either or both parties. It is not at all unusual to see older lizards in nature bearing the scars of past aggressive encounters, either on their heads or elsewhere on their bodies. Other injuries that my colleagues and I have observed in various lizard species that probably result from male combat include broken jaws, broken tails, and missing digits. No wonder lizards strive to avoid violent confrontations wherever possible!

Despite this wholly logical argument for escalated and unrestrained male combat being a Very Bad Idea, some animals nonetheless fight in exactly this way. Always, though, there is a good reason. The aptly named gladiator frogs in Central America spawn in ephemeral pools of water that form after heavy rains and disappear soon after. This gives frogs an incredibly short window of time to find a mate before their chances of reproduction literally dry up. Female gladiator frogs have to lay their eggs in the water and are necessarily drawn to these short-lived water bodies, creating scattered clumps of females centered around transient pools. Males thus have an opportunity to greatly increase their reproductive success relative to other males by taking control of a pool and keeping all other males away from the females therein, and this is what they attempt to do. But because there is so little time for breeding, there is none for the niceties of display and ritualized combat—the males need to mate, and they need to mate *now*, because if they do not do so there is no guarantee that they will survive until the next rains bring another mating opportunity. Males therefore fight to the death over access to these pools and the females they contain, using bony spurs protruding from their forearms as daggers in batrachian knife-fighting duels with everything at stake. (Yet another frog, often referred to as the Wolverine frog, produces claws by breaking the tips of its own fingers such that the sharp broken bones pierce the flesh.)

Homicidal frogs are the extreme end of a male combat continuum, with the other end being species such as butterflies where males simply display, flittering around aggressively but never touching one another. Between these two extremes are all manner of degrees of physical ferocity, from aggressive lizards to cuttlefish, which escalate from initial displays involving rapid changes in coloration and patterns to grappling and savage biting. All of these fights, however, be they knock-down, drag-out brawls or the animal equivalent of belligerent Morris dancing, are enabled by animal athletic abilities, albeit in some cases more directly than in others.

Flaunt It If You've Got It

Because most animals that are not gladiator frogs prefer to avoid injury wherever possible, the escalating nature of male combat depends on the initial round of low-level signaling. Indeed, the commonly accepted explanation for these signals and displays is that they convey something important about the animal that is relevant to his ability to win a fight. The traditional notion is that displays advertise information on body size, because of the strong effect of body size on male combat outcomes. Bigger animals are just more likely to win a fight against a smaller opponent, which is why we humans have weight divisions in boxing and other combat sports. This size effect is so overwhelming that the other male is liable to back down and not engage if you can just convince him that you are bigger than he is, hence the weird lateral compression display and accompanying side-on posture of the collared lizards described above. Escalation of combat to the third and final stage in lizards—direct physical confrontation through biting—tends to occur only in similarly sized individuals, because until this point the animals have not been able to figure out what differences in fighting ability, if any, exist between them. They therefore have no choice but to resolve things directly.

Once combat has progressed to this point, however, winning a fight cannot be just about body size. After all, if two males of exactly the same size enter into a fight, there has to be some way of determining a winner and a loser. In nature, unlike in human sports, these fights cannot end in a draw— the fitness stakes are too high—so there have to be other things that make up fighting ability besides size alone. More and more, we are learning that the true predictors of animal fight outcomes, as in collared lizards, involve various kinds of whole-organism performance.

If this is the case, then signals should have evolved to convey information on athletic abilities that influence fight outcomes instead of, or in addition to, information on body size. But whereas lizards can directly display their jaw muscles for other lizards to gaze on and tremble, that only works when the two rivals are reasonably near each other. Males strive to keep their distance if at all possible, so it would be useful to have a system for advertising one's fighting ability to rivals before getting close enough to engage physically. In this way, males could gauge each other's martial skills and assess whether starting a fight is really a good idea. For males that have territories and females, such long-range advertisement signals could also act as a deterrent, convincing rivals that they are better off trying their luck elsewhere. What this means is that there has to be some characteristic of an aggressive signal or display that is correlated with an individual's fighting ability. And just as important, that specific characteristic should be independent of the general relation between the signal and body size.

This property of signals as performance advertisements is exemplified by the dewlaps of some species of lizards belonging to the genus (guess which one!) *Anolis*. As I've mentioned, anoles are remarkably diverse in morphology, ecology, and behavior. One example of this diversity is in the apparent variety of mating strategies. Anoles appear to exist along a continuum of territoriality, with some very territorial species that aggressively exclude males of the same species from their territories and other species that are not especially territorial and are more tolerant of having other males around. Escalated male combat in anoles involves biting, just as it does in collared lizards and in many other lizard species. But anoles are also noteworthy for being very visual lizards, and males defending territories display conspicuously. Displays in male anoles involve the dewlap, an extensible throat-fan that varies greatly in size, shape, and coloration across species. Although dewlaps are employed in various ecological contexts, ranging from courtship of females to territorial displays, they are used most obviously during aggressive interactions with other males. Anole dewlaps are colorful, beautiful, and, above all, mysterious.

Despite male anoles' readiness to deploy their dewlaps, researchers have had a hard time figuring out exactly what it is that dewlaps do. How they are used, as well as the variation in their design throughout *Anolis*, suggests that they are a signal—but of what? One intriguing possibility is that they are signals of performance, specifically bite performance. Evidence for this idea comes from work by Bieke Vanhooydonck of the University of Antwerp and colleagues who showed that bite force is positively correlated with dewlap size in several species of anoles independent of body size. This caveat is important, because we might expect many characteristics of animals to be positively linked to each other simply because both are also positively linked to body size through a phenomenon known as allometry, or scaling. Bigger animals have both larger dewlaps and higher bite forces than smaller ones, so we shouldn't be very surprised that large animals have both large dewlaps and bite forces. However, it turns out that if you remove these size effects through statistical wizardry, dewlaps and bite force are still positively related

Fig. 3.1. Male *Anolis cristatellus* displaying his dewlap.
Dewlap size appears to signal bite force in this species, as
in several others. Photo by Michele A. Johnson.

to each other. This means that animals that can bite hard *for their size* also
have relatively larger dewlaps *for their size* (fig. 3.1).

This interpretation of dewlaps as a signal of bite force makes sense because if you set up Ultimate Lizard Cage Fights between same-size males
in these particular territorial anole species (as I did for part of my doctoral
dissertation), the winners overwhelmingly tend to be those males who can
bite hard for their size. Anole dewlaps therefore appear to signal at least two
things to rival males: their body size, through the absolute size of the dewlap, which is positively linked with body size, and their bite force, through
their relative dewlap size, which is linked with bite force independent of
scaling effects.[2]

The evidence for anole dewlaps as performance signals does not end
there. Among Caribbean *Anolis* as a whole, dewlaps and bite forces are

2. I hasten to add that there is far more to anole dewlaps than this relation with bite
   force, and dewlaps are almost certainly affected by a number of other selective pressures in various contexts. It is also worth noting that this refers only to dewlap size and
   that other aspects of the dewlap, such as color and patterning, show different trends.

related to each other in this way in territorial species where males are observed to fight a lot, but they are not related in less territorial anoles that fight less frequently. Furthermore, whereas males that bite hard for their size win fights most often in territorial species, bite force does not affect fight outcomes in those less cantankerous non-territorial species. These results imply not only that territorial males signal their bite forces to each other through the relative size of their dewlaps, but that non-territorial species do not do this, presumably because male combat (and thus biting) is less important to them.

Anoles are not the only lizards that use visual signals to advertise performance abilities used during male combat. Territorial male tree lizards, *Urosaurus ornatus*, possess a colored throat patch not unlike anole dewlaps. But they also have colorful patches on their bellies that they display to other males during aggressive interactions. Unlike in anoles, the size of the tree lizard throat patch does not appear to convey any information on biting or other whole-organism performance capacities to rival males (although the throat patches are meaningful in other ways). However, the size of the colored belly patch is linked with bite force. This suggests that animals with complex patterns of ornamentation are probably signaling a variety of things to other individuals besides performance, although performance is certainly in the mix.

Combative Crabs and Battling Beetles

Lizards aren't only great organisms for studying performance, as noted previously, but male combat is also common among lizard species. This makes lizards the perfect animals for scientists such as myself interested in the intersection between male combat and performance, and researchers have therefore paid them a lot of attention. But this raises important questions. Isn't it possible that this relation between performance and signaling is lizard-specific and that other types of animals resolve their differences in some other way? What about other animal species that might rely on performance capacities other than bite force to win fights? And do animal signals ever advertise other performance abilities, such as endurance capacity or sprint speed? Even in lizards, bite force is not the only performance ability that has been linked to fitness—in collared lizards, for example, number of offspring is also predicted by sprint speed, with faster males siring more offspring than slower ones, probably because they can place themselves and their jaws in the path of an invading male faster than a slower lizard could.

There is evidence from a growing number of studies in a variety of animal species that the signals and displays used by males in aggressive interactions are often linked to performance abilities that aid individuals in winning fights. An example of such a link in a nonlizard comes from hermit crabs, which fight not over access to females, but over ownership of shells. Hermit crabs do not produce their own shells, and are forced to find and inhabit the cast-offs or remains of other shell-bearing animals, such as snails. Such discarded shells are a limited resource, however, and whenever you have a limited resource that is in demand by an entire group of animals, you have conflict. So scarce are uninhabited gastropod shells that crabs may well resort to taking them away from other animals that already have them.

This is the strategy employed by the hermit crab *Pagurus benhardus*, which is not above rising up against the shell-bearing elite and grabbing some of that sweet snail-shell action for itself. If one crab decides that it is hard done by shellwise and covets the shell of another crab (perhaps because it is outgrowing its current accommodations), it will try to take it. Ritualized combat in these creatures takes the form of shell rapping, whereby an attacking crab grabs onto the shell that it wants with its legs and bangs its own shell against it. The combatants rap back and forth in this manner until such time as the dispute is settled. Victory in these fights goes to the animal that can rap harder and for longer than its rival. The ability to do so is correlated with how strong a particular crab is, which is effected by its muscularity. Victorious crabs therefore have more abdominal muscle in particular compared to losers, and also (perhaps incidentally) show greater sprint speeds and endurance capacities compared to losers. Shell rapping appears to signal strength and overall performance to rival crabs, one of which usually gives up if it decides that it is outmatched. If the loser is the defender, that animal either abandons the fight and leaves its shell voluntarily or, if it is particularly stubborn, is forcibly evicted from its shell by the other animal.

The dung beetle *Euoniticellus intermedius* also appears to use a performance-based combat strategy. At a maximum adult length of less than 1 cm (0.4 in.), an angry *E. intermedius* male may not seem too intimidating. Yet these little beetles are fighters, and they possess a specialized weapon that they use during aggressive interactions with other males. This weapon is an outgrowth of the beetle's exoskeleton that forms a horn, resembling a rhino horn in relative size and placement, and is just one of many types of horns wielded by dung beetles the world over. Dung beetles are perhaps the embodiment of male combat in the animal kingdom, and various species use their horns, which range from the rhino-type nose horn of *E. interme- dius* to the aurochs-like horn of *Onthophagus taurus* and the triceratops- style horns of *O. haagi* to the spectacular, swept-back antler-horns of *O. ran- gifer*, to gain leverage and tip each other over during fights over access to narrow underground tunnels containing the female of the species and a ball of dung within which she will lay her eggs.

Defending the tunnel is a good idea for these guarder males, because if they can keep rivals away from the female and her brood ball within, they can monopolize her eggs and thereby increase their fitness. Sometimes fights between rival males even occur within the narrow tunnels where, with little room to maneuver, males use their horns either to push their rival out of the tunnel or to shove him aside and attempt to pass him in the tunnel. In a species like *E. intermedius*, where the horn is sort of stubby rather than sharp (fig. 3.2), it's hard to imagine how males could injure each other in this way, but these fights can nonetheless be energetically expensive if they go on for too long. Again, it is in everyone's own interests to find a way to gauge each other's fighting ability before engaging in a long and costly bout.

In staged competitions between *E. intermedius* males, horn size is the most important predictor of victory, particularly in larger males. Just as in anoles, the size of the horn is also positively linked (independent of body size) with two performance capacities that are useful in a dung beetle fight, namely strength, measured as the ability to resist being pushed or pulled out of a tunnel, and exertion. Even in the dark, dung-laden tunnels within

Fig. 3.2. *Euoniticellus intermedius* males:
horny. Photo by Rob Knell.

which these creatures fight, there is ample opportunity for assessment of a rival male's horn size, and the existing evidence is consistent with the notion that horn size is an *honest signal* of performance abilities relevant to fight outcomes in *E. intermedius,* just as dewlaps are in territorial anoles. The existence of such similar performance signals in animals that are so distantly related to one another suggests that honest performance signals could be widespread throughout the animal kingdom and thus a common feature of male combat.

Never Trust a Crustacean

These signals, and others like them, are considered to be honest because they reliably convey information on individual characteristics to other males. An important characteristic of such a signal is that it is difficult to fake. In some species, honesty is mandated by physical limitations of the signaling apparatus. Male red deer, for example, roar at one another, and the pitch of that roar is constrained by the size of the larynx such that only very large males can produce low-frequency calls. Thus, roar frequency is an honest signal of body size in red deer—a buck cannot fake a deep call, thereby misleading other deer as to its size. In other animals, honesty is probably upheld by the various costs of expressing (or building) and maintaining signals, which are such that only certain animals can express those signals without incurring significant penalties.

Be that as it may, many of you may be narrowing your eyes suspiciously at all of this talk of advertisement and honesty. Your skepticism is more than justified. Human advertisers practice deception because it is in their own best interests and is a tried-and-true method of separating consumers from their money. Indeed, the prevalence of misrepresentation, distortions of the truth, and outright fabrications in human advertising are such that the savvy consumer would do well to take any claims made by an advertiser with a truckload of salt. So why would advertising in the animal world be any different? What would be the individual benefit of signaling to a competitor exactly what you are capable of, when a far better strategy would be to exaggerate or conceal your abilities? Consider the sequential assessment game. It starts off with potential combatants signaling or displaying with the aim of convincing each other to disengage and admit defeat without resorting to fisticuffs. Wouldn't it make sense to lie about your own fighting prowess, thereby convincing your rivals that you are some kind of fighting machine and intimidating them into backing off?

There are several reasons why this kind of deception is unlikely to be rampant in nature, but they all boil down to the various and substantial costs of being exposed as a cheater. Signaling dishonestly is risky; but this is not to say that animals never bluff or hoodwink one another. One of the inherent problems with studying dishonest signaling is that it is, by definition, dishonest. Such signals are intended to go undetected, which means that picking up on them is a challenge. We nonetheless have reason to believe that some animals do lie in exactly this way, and work on crustaceans has revealed cases of disingenuous signaling during combat.

Fiddler crabs are charismatic little creatures that make frequent use of visual signals involving the male claw. In males, one of the two major claws—either the left or the right—is greatly enlarged relative to the other. The smaller claw is used for feeding and all of the other regular things that crabs use their claws for. But the larger, major claw is special. Not only is the major claw bigger, accounting for a third to a half of the crab's body mass, but it is often colorful, with different species having claws of varying sizes, shapes, and colors. Fiddler crabs derive their name from the way that males use their major claws during signaling: they hold them out in front of them and wave them up and down, creating a motion not unlike that of a fiddle player. The wave display of the fiddler crab is one of nature's true delights, especially when multiple crabs display simultaneously, with competition for being the first to display to females resulting in near-synchronized waving that makes the mud flats these crabs inhabit seem alive with colorful motion.

Major claws are not only used for signaling to females, however; they are also used for combat. Male combat in these crabs involves an initial round of signaling during which the combatants compare and assess the sizes of each other's major claws. If assessment of claw size does not settle a dispute, the interaction escalates in the same way that lizard fights do, ultimately resulting in physical combat where males may use their claws to wrestle each other into submission. Pat Backwell at the Australian National University in Canberra has done excellent work on fiddler crab behavior and ecology, and her lab was the logical choice to work with when I became curious about crab combat, and especially the strength of their claws. In the fiddler species *Uca mjoebergi*, which lives on the mud flats around Darwin in Australia's Northern Territory, claw size predicts at least two types of performance abilities independent of body size: claw pinching force measured with a bite-force meter (and analogous to bite force, except with claws) and, as in *E. intermedius* (and for the same reason), the ability to resist being pulled out of a tunnel (fig. 3.3). Rival males can therefore accurately glean information on both of those performance capacities from relative claw size and then use that information to decide whether to escalate an interaction, just as territorial anoles do. However, the situation gets more complicated once we realize that these crabs can do a very unusual thing: they can lose their major claws and grow new ones.

The ability to regenerate lost body parts is widespread in the animal kingdom in one form or another. Some lizards, for example, can voluntarily drop their tails (a process called *autotomy*) and usually do so as an antipreda-

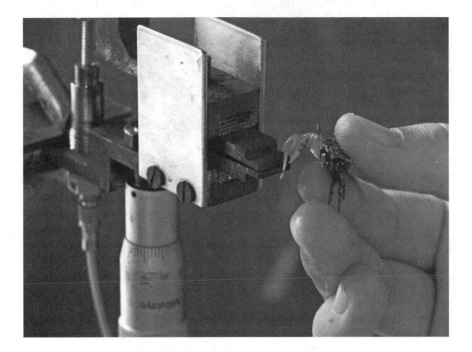

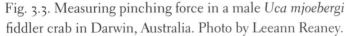

Fig. 3.3. Measuring pinching force in a male *Uca mjoebergi*
fiddler crab in Darwin, Australia. Photo by Leeann Reaney.

tory mechanism. The lost tail can be regrown, albeit in altered form, with
the replacement tail supported by cartilage rather than bone. Salamanders
famously can regrow entire limbs, which lizards cannot do.[3] Fiddler crabs
can regenerate, too, and if they happen to misplace a major claw, perhaps
in a close encounter with a predator or during a fight with another male,
they can grow another. The regenerated claw, however, is subtly different to
the original; it is lighter and slimmer, has less area for attachment of muscle
to power claw closing, and lacks tubercles—small bumps or nodules on the

3. The regenerative abilities of lizards are remarkable, and some species can regenerate
skin, the optic nerve, and even parts of the brain! Nonetheless, as a longtime reader
of Spider-Man comics, I have always been puzzled as to why Dr. Curt Connors (aka
the Lizard) studied lizards to figure out how to regrow his lost arm, when he really
should have been working on salamanders. Connors ultimately managed to succeed
with lizard DNA anyway, so I suppose Stan Lee is a better biologist than I am.

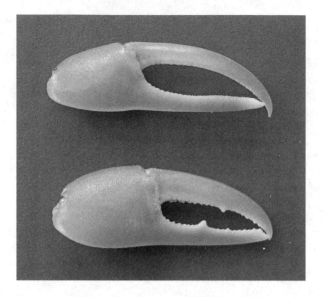

Fig. 3.4. Regenerated (top) and original (bottom) *Uca
mjoebergi* fiddler crab claws. Only the original claws
predict performance. Photo by Tanya Detto.

inner part of the claw-closing surface (fig. 3.4). But whereas we can look at
an original and a regenerated claw and clearly differentiate one from another based on the details listed above, male fiddler crabs assess claws based
on other criteria, such as length, that are superficially similar in original and
regenerated claws. This observation, in turn, explains another surprising
find: male crabs cannot tell the difference between original and regenerated claws.

The male crabs' inability to distinguish between original and regenerated claws means that they can be fooled by males wielding regrown claws,
because whereas the original claws convey accurate information on individual performance capacities in U. *mjoebergi*, regenerated claws do not.
In fact, there is no predictive relation at all between claw size and either
pinching force or pull-resisting force in regenerated claws as there is for
original claws. This remarkable situation allows males with regenerated
claws to deter other males (who may actually be better fighters than they)

from engaging them in escalated combat, essentially bluffing those better-performing males into backing down from a fight. Regenerated claws are therefore a fraud perpetrated on other male fiddler crabs, one that takes advantage of an apparent glitch in the crab's visual system to intimidate rivals by advertising performance prowess greater than that which the bearers possess in reality.

Data from natural populations harboring males with regenerated claws show that this bluff works, but it has limits. Male fiddler crabs fight over territories containing females, just as many other animals do, and the first obstacle stems from the differences between taking a territory away from someone and defending a territory of one's own. Male *U. mjoebergi* crabs with regenerated claws who challenge territory holders can indeed bluff males into giving up their territories without a problem; however, when they themselves hold a territory and are challenged by another male, they are more likely to lose it. This is because wandering regenerated-claw males without a territory have the luxury of choosing their opponents, whereas regenerated-claw males defending a territory have no choice but to take on every male who tries to take their territory away. Eventually they will have to face a male who won't back down, meaning that that fight will inexorably escalate to physical combat, exposing the deceptive male's sham.

The second problem is that bluffing, like many other things in life, can become a victim of its own popularity. The phenomenon at play here is called *negative frequency-dependence*. This means that bluffing is effective if few individuals in a population do it, but that effectiveness rapidly drops off as more and more individuals act deceptively. As an analogy, consider a poker game where a player chooses to bluff every hand, regardless of the hand that player is dealt. The other players will soon cotton on to this strategy and call that bluff every time. On the other hand, bluffing at a very low frequency just might work. This trade-off in bluffing frequency also exists in fiddler crabs, albeit across the population as opposed to individually.

In the Darwin *U. mjoerbergi* population, the frequency of deceptive males with regenerated claws is about 7 percent, which means that lying males are just rare enough to avoid raising the suspicions of other crabs. However, the proportion of dishonestly signaling crabs among populations is variable, depending on how many individuals have found it necessary to regrow their claws. For example, in the related southern African fiddler crab U. *annulipes*, frequencies of regenerated claws in three populations are much higher, ranging from 16 percent to a surprising 44 percent. For populations containing large percentages of dishonest males, it should no longer pay crabs to trust claw size as a signal of performance, and one would therefore predict that fights will escalate more readily in populations with high frequencies of regenerated claws.

This has yet to be shown conclusively, but evidence suggests that this is what happens. For example, among ten populations of the fiddler crab U. *vomeris* (studied by Candice Bywater during her doctoral research at the University of Queensland), with percentages of males with regenerated claws ranging from 2 percent to 35 percent, those populations with higher frequencies of dishonestly signaling males also show higher rates of aggressive interactions. The work on U. *annulipes* also turns up the intuitive finding that populations with large numbers of older and larger crabs exhibited higher frequencies of males with regenerated claws. This is because older individuals are more likely than younger ones to have lost a claw at some point in their lives. These crabs are not known to self-amputate their claws, which raises the intriguing possibility that the dishonesty is an opportunistic strategy used by individuals when the demographic make-up of the population allows for it. If so, a dishonest system that works perfectly well in a given population could stop working entirely if the age structure, predation pressure, or density changes unfavorably for whatever reason.

A second, similar example of dishonest signaling comes from another type of crustacean: the crayfish *Cherax dispar*. Male crayfish are both territorial and aggressive and use their two enlarged front claws for intimidation and combat just as male fiddler crabs use their one major claw. Also much like male fiddlers, relatively few of these interactions escalate to direct fighting: more than 80 percent of fights are resolved without individuals engaging in physical combat, and claw size is the most important predictor of male combat outcomes, again suggesting that rival males are inferring information regarding fighting ability from those claw displays. However, claw size is not a reliable indicator of claw strength in C. *dispar*, and measurement of the muscle inside male claws shows that it surprisingly weak. In fact, male claw muscle is only half as strong as the claw muscle of females, which do show a predictive relation between claw size and claw strength! This evidence suggests a mechanism underlying claw dishonesty whereby some males—perhaps especially males that lose their claws and grow new ones in the same way that male fiddler crabs do—invest more in growing large claws than they do in building expensive, high-quality claw muscle and are thus able to score energetic savings while still bluffing other males into backing down from fights before their deception is exposed.

It is almost certainly not a coincidence that our two best understood examples of dishonest signaling are found in crustaceans. Unlike in vertebrates, where muscles are attached to and surround an internal skeleton and are thus readily apparent to any who cares to admire that vertebrate's muscularity and draw inferences regarding its athletic potential, crustacean muscle is hidden from view within a hard exoskeleton. Crustaceans also tend to have large, functional claws, and the size of the claw muscles in particular can only be inferred from the size of the claw itself. It is thus easy for a crustacean to hide the strength of its claws by developing claws that are larger than they are muscular. Robbie Wilson, the mastermind behind the crayfish research program, summed up this situation in a statement that, much like Wilson himself, would appear simplistic if it weren't also so insightful: "It's easy to be dishonest when you're hiding what's inside."

Are Athletes Sexy?

Performance capacities can and do underlie both the process of male combat and the signals involved in it. But what of female choice? Whereas it is intuitive that such displays would signal aspects of individual performance abilities to rivals that males are apt to use those performance abilities on, it is far less clear whether males would signal their performance abilities to females or indeed whether females would even care.

In many species, males gather in a certain area and simultaneously display to females. These gatherings of males are called leks, which is derived from a Swedish word that describes playing children's games. Those areas harboring leks are referred to as arenas, and so it would be perfect if the purpose of these arenas were to show off athletic abilities to females. But nature has no respect for an elegant metaphor, and the evidence that females choose males based on their athletic prowess is mixed. Part of the problem is that it is unclear exactly what a female might gain from mating with a high-performance male.

There are two kinds of benefits that females could potentially expect from mating with a particular male. *Direct benefits* accrue to the female herself and come about because the male (or males) whom she mates with provides her with resources, protection, or assistance with parental care. This is in contrast to *indirect benefits*, which are benefits that accrue, not to the female, but rather to her offspring. For example, if a female mates with a male possessing certain qualities—say, exceptional stamina or speed—and that male does not stick around to look after her or assist with parental care duties, then the only advantage the female gets from mating with that individual is that her offspring might inherit that same heightened performance from their father.[4] Given the positive relation between performance and fitness in animals such as collared lizards, those indirect benefits could be large. A male that is a good performer and who is able to provide females with captured prey or protection would be very useful to a female. But since males are like unto dust in the wind immediately following copulation in the vast majority of animal species, indirect performance benefits (if any) are probably more important than direct ones.

Studies testing the idea that females might prefer good performers specifically are thin on the ground, and the few that do exist are contradictory. For example, female crossbill finches prefer to mate with males who are fast foragers, and they probably gain direct resource benefits from doing so. But female green anole lizards, who would gain only indirect benefits from mating with high-performance males, show no preference for males who can jump or bite especially well. Males are similarly poor parents in fish

---

4. It is also possible that females who mate with attractive males could gain indirect benefits in terms of attractive sons. These two ideas, namely "good genes" and "sexy son" benefits, respectively, have a complicated and controversial history in sexual selection that is far beyond the scope of this book.

such as the pacific blue-eye, yet females are attracted to blue-eye males with long dorsal fins. Dorsal fin length is positively related to swimming speed, strongly suggesting that females exhibit either a deliberate or incidental preference for good swimmers. Likewise, female guppies are attracted to deadbeat males with particular combinations of colored patches on their body, and some of those attractive color combinations are also associated with superior swimming performance in male guppies.

Encouraging as these findings for fish are, the test for indirect swimming performance benefits doesn't hold up, at least in guppies; the offspring of attractive male guppies are not better swimmers than those of unattractive males. The reasons for this finding are potentially complex, and work on a very different animal offers hints as to what they might be.

In the Australian black field cricket (*Teleogryllus commodus*), male attractiveness is determined by *calling effort* (how often males call) — males who call a lot are more attractive to females than males who call less. As far as performance goes, a given male's ability to win a fight depends in part on his jumping ability. Good jumpers win more fights than poor jumpers, in combination with other traits such as size and biting ability. So good jumping abilities are useful for males, and you will not be surprised to learn that the *T. commodus* male loses interest in the female once he is sure that she has been properly inseminated. Females might well gain indirect benefits in terms of high-performance sons by mating with high-performance males; however, breeding experiments aimed at revealing the genetic relations among these traits in *T. commodus* show that both calling effort and attractiveness have a negative genetic relation with jumping ability. This means that if a female chooses to mate with an attractive male who calls a lot, her offspring will inevitably be terrible jumpers. In fact, because of the nature of the genetic relations among these traits, it is not possible for a male to be both attractive and a good jumper, and any female who mates with an attractive male incurs an indirect genetic *cost* in terms of unathletic offspring!

At least, this is true most of the time, and herein lies the rub. These same experiments also yielded evidence for two other, different genetic combinations of attractiveness, calling effort, and jumping ability. Plainly stated, there are two other ways to make an attractive male cricket that don't necessarily involve being bad at jumping (although these other combinations are almost certainly far less important than the main attractiveness-jumping trade-off). This means that females may indeed gain indirect performance benefits from mating with athletic males—but only rarely. The patterns of female selection on male guppy ornaments are similarly complex, and work on guppies aimed at understanding what female guppies really look for in a male shows that there are also multiple ways to be an attractive guppy. Those aspects of coloration that reflect swimming performance could therefore be implicated in attractiveness in various ways, albeit with some of those ways again being more important than others. The potential indirect benefits to females of mating with athletic males are therefore neither straightforward nor necessarily universal, and teasing apart the genetic relations among different types of performance, attractiveness, and male ornamentation and display remains a significant challenge.

Further tentative evidence for female choice for good performers comes from perhaps the last animal you might expect—human beings. Erik Postma of the University of Exeter became interested in these questions about female preferences and performance and went about testing them in a crafty way. He asked more than eight hundred women to rate pictures of the faces of eighty competitors in the 2012 Tour de France cycling race for attractiveness, masculinity, and likability and then analyzed the results while statistically accounting for a number of potentially important factors. Although the raters did not know the cyclists' times, Postma found that riders with better times in the Tour de France were nonetheless also rated as more attractive, and this preference for cycling performance was strongest in women not using hormonal contraceptives. Erik also found that attractive cyclists were considered to be more likable, a result that is probably explained once you consider that the raters didn't meet the cyclists in person and therefore were not subjected to interminable conversations about cycling. Given the complexity of human mating decisions (Postma did not note whether any of the raters were drunk at the time), it is unlikely that performance is the only thing that raters were focusing on. Nonetheless, these findings raise intriguing questions as to how endurance performance might be reflected in the facial features of male cyclists and indeed the displays of other animals.

FOUR

Girls and Boys

Males and females are different. Considered throughout the animal kingdom as a whole, these differences between the sexes are frequently striking, often puzzling, and always interesting. Depending on the species in question, males and females might differ in size, shape, color, behavior, or all of the above. In some cases, sex differences are so extreme that males and females bear little to no resemblance to each other, and it is not unheard of for taxonomists to assign males and females from a given species to different species altogether based on how they look.

Possibly the most extreme example of this *sexual dimorphism* is that of the anglerfish, a deep-sea dweller in almost total darkness. In this bizarre animal, females assume a regular fishlike shape, albeit replete with the obligatory array of terrifying teeth and outlandish features that characterize deep-sea organisms. The males, however, are about as different from the females as one could imagine. Male anglerfish not only don't resemble females at all but don't even resemble fish, being little more than tiny, testes-containing bags of flesh. Like many other animals that are very small or live in perpetual, almost total darkness, male anglerfish face significant hurdles in locating a female to mate with. If by some miracle a male does find one, he isn't likely to encounter another anytime soon or, perhaps, ever. Male anglerfish are therefore the ultimate stage 5 clingers, insofar as they attach themselves to the first female that comes along and they never let go. Such is the extreme sexual dimorphism in anglerfish size and shape that male anglerfish were originally thought to be parasites feeding off the female, and in a sense they are, as each lucky male fuses permanently with his female and lives off the nutrients in her bloodstream. But they are also the male of the species, and a mate that will never, ever leave.

Size differences between males and females are common in animals, and because of the effects of size on performance and athletic ability, these differences have clear implications for function. In humans, for example, males are on average taller, heavier, more muscular, and shaped differently than females. These differences, as well as the physiological factors underlying them, collectively account for the many differences in athletic ability between men and women. Men are 5 percent to 10 percent faster than women over short-, middle-, and long-distance athletic events, and world records for short-distance track performance events in particular have been, and continue to be, markedly faster for men than for women. Improvements in training, nutrition, and other factors over the years breed continuous speculation that these differences might ultimately disappear, and curves such as those shown in figure 4.1 are often extrapolated—a risky business in statistical analysis—to suggest that men and women will converge in track performance at some point in the future. This is plausible (though still dubious) only for long- and ultra-long-distance endurance events such as the marathon (fig. 4.1d). For many other athletic events, such convergence is extremely unlikely unless male and female athletes also ultimately converge in size, shape, and physiology.

But to understand the implications and origins of sex differences in performance and the factors affecting performance, we first have to know something about why the sexes differ at all. In the previous chapter I talked about sexual selection and how males often compete for females. Now I'll briefly summarize the reasons why this happens, because we have to deal with these if we are to really understand why males and females are different (note that that this enormous and important topic could easily be the subject of its own book, and its treatment here is necessarily superficial).

In almost all cases, sex differences and sexual selection stem from the nature of sexual reproduction. Specifically, they are either the direct or indirect consequence of a phenomenon called *anisogamy*. This means that males produce lots and lots of tiny sperm over their lifetimes whereas females produce relatively few, comparatively large eggs. Thus, even though the

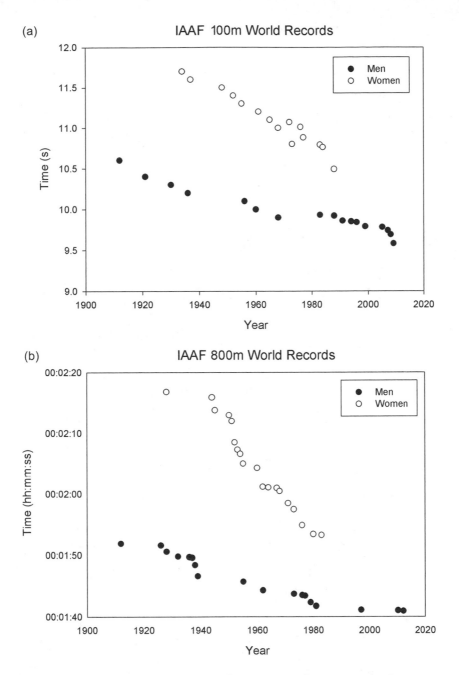

Fig. 4.1. Progression of official International Association of Athletics Federations world records over time for men and women for (a) 100 m, (b) 800 m, (c) 1,500 m, and (d) marathon. The weirdness in the times around the mid-1970s for (a) is due to the switch to more accurate electronic timing in 1975. For years where multiple records were set, only the fastest times are shown.

(figure continues)

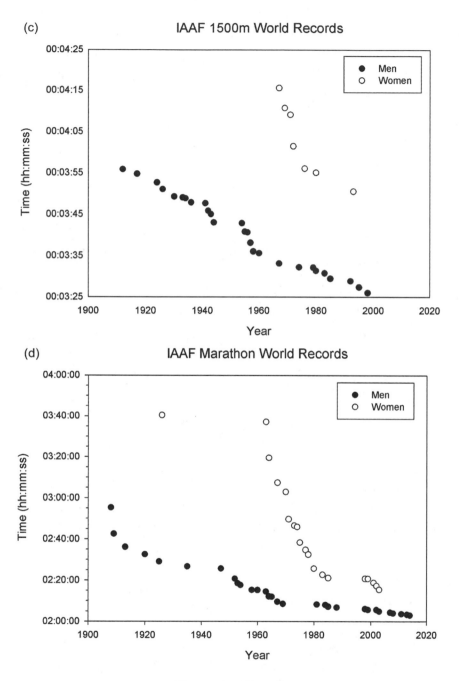

Fig. 4.1. (*continued*)

total lifetime energetic reproductive investment in gametes (reproductive cells) of males and females may be equal, for any given bout of reproduction between a male and a female, females invest more and thus have far more at stake in each reproductive event than do males. It is the female, after all, that has to, at a minimum, carry the fertilized eggs and who often invests significant energetic resources in those eggs even if she doesn't provide parental care to the offspring (which she might also do).

This simple fact has enormous consequences and drives a number of general trends in animal mating systems. One of the most important insights to result from the consideration of sex-specific reproductive investment is the notion of a *Bateman gradient*. A Bateman gradient describes the relation between the number of matings and the number of offspring within each sex, and it derives its name from experiments on fruit flies done in the 1940s by Angus John Bateman. What Bateman showed, and has subsequently been borne out in investigations of the mating systems of many other animal species, is that males usually exhibit a positive Bateman gradient, which means that males who have more mates sire more offspring. This makes sense—if a male has sex with more females, he can potentially impregnate more of them. But for females the Bateman gradient is generally flat, which means that females get no advantage (in terms of *number* of offspring per mating) from mating with lots of different males. Indeed, one male ejaculate is usually sufficient to fertilize all of her eggs at any given time. Females simply do not benefit from having sex with lots of individuals in the same way that males do (although if multiple males of variable genetic quality fertilize her eggs, she may get other, indirect benefits in terms of better offspring as opposed to more of them, as discussed in Chapter 3).

This means that males can and do spread their seed with abandon, ejaculating everywhere and in anything, whereas females must make more prudent reproductive decisions.[1] This observation also explains why sexual selection tends to operate more strongly on males than on females. Males

1. This is not to imply that males pay no costs of reproduction; they do, and those costs can be high indeed. But the costs of reproduction for males often arise from protracted or expensive attempts to attract and mate with females, and they seldom manifest as direct energetic investment in the offspring themselves.

need to compete with other males for females, because males who have sex with more females have more offspring. Females, whose eggs are the limiting resource, get control over whom they allow access to that resource.

Anisogamy, and the consequent fact that females invest more than males in a given bout of reproduction, has huge implications for how males and females go about their daily lives. Many of these implications were first recognized by Robert Trivers in the 1970s, and his papers on the subject constitute the most influential work on evolutionary biology since Darwin. For one thing, differences in investment can affect the size and shape of one sex relative to the other. In some cases, females might be larger than males because female size is directly linked to fecundity or reproductive potential (this is because absolutely larger animals have more resources immediately available to invest in their offspring than smaller ones). Alternatively, it could drive the evolution of larger males because males, in their tireless quest to inseminate everything, need to be large to fend off other males and gain access to females. The fact of larger female reproductive investment also means that females get to be choosy about their mates, driving males to bear the costs of display and constant seeking of female attention, which larger individuals are usually in a better position to do. In general, insects follow the former pattern, whereas reptiles and mammals favor the latter, although this is by no means a rule and there are many, many exceptions.

Whatever the direction of sexual size dimorphism, sex differences in size and shape, and the tendency for females to bear the brunt of the energetic costs of reproduction in terms of carrying eggs or offspring around with them, certainly affect their athletic abilities. As we shall see, though, it doesn't always do so in the ways that we might expect.

Why Not to Be a Male Spider

Spiders might well be considered the poster child for sexual dimorphism. Web-building spiders in particular exist along a continuum of sexual size dimorphism, ranging from little to significant sex differences in size. But in almost all cases where dimorphism exists in spiders it is in the direction of larger females—sometimes much larger females.

This large discrepancy in size creates several unusual problems for the males of those very dimorphic species. Chief among these is that, for males, spider sex is a very risky activity. Because the female might be an order of magnitude or more larger than the male, and because there is little advantage to her in mating with lots of males, she will often quite happily eat every small living thing she finds—including male spiders. That she might well be consuming a male of the same species as herself, a potential mate and possible father of her children, bothers her not a whit. To her, males are small, snack-sized things, and she needs to eat. There are enough males around, however, that one of them will likely escape her attention at some point, possibly while she is busy eating some other male, and then she will be inseminated and her eggs fertilized. How the males go about it is no concern of hers. It's up to them to find her, not the other way around.

Therefore, sex from the perspective of a male belonging to an extremely female-biased dimorphic species of spider looks something like this: first, he has to find a way to get close to this other, gigantic spider that would just as soon eat as look at him; then he has to copulate with her; then he has to survive copulation; and last, he has to get away from the monstrous female spider that he has just copulated with, again without being eaten. This is such a tall order that some males never achieve the final two steps at all. But males of some species do, and they have evolved a variety of strategies to achieve just this.

The male desert spider essentially date rapes the female, spraying her with a chemical he himself produces that renders her immobilized long enough to allow him to have his way with her. Males of Darwin's bark spider on Madagascar, for their part, do something that females reportedly consider far more appealing—the male copulates with the female while she is molting and thus relatively harmless, and their courtship involves the male "lubricating and nibbling on the female's genitalia," which, according to an interview (that I did not make up) in *New Scientist* with researchers studying this species, "relaxes" her and makes her less likely to cannibalize him. For male black widow spiders, by contrast, their cause is hopeless, and they make no attempt to avoid becoming a meal for the larger female. On the contrary, the black widow male does something quite surprising; following copulation, he performs a vaulting maneuver over the female that would make any Olympic gymnast proud and ends with him landing directly on the female's fangs! This cannot be interpreted as anything other than direct self-sacrifice. It likely represents the most extreme act of parental investment on the planet, with the male providing food resources to the female and, ultimately, to the eggs she will soon lay—eggs that he himself has just fertilized.

As cool as suicidal vaulting male black widows are, though, they aren't why I wanted to talk about spiders in the context of sex differences in athletic ability. Rather, I wanted to bring up a different species: the cobweb spider *Tidarren sysyphoides*. Male cobweb spiders are fortunate insofar as their females are far less prone to cannibalism than other spiders. Instead, they face a different problem. Males are roughly 1 percent of the mass of the females, very small indeed. Consider, for example, an average-sized man attempting to have sex with a woman a hundred times his size,[2] and you get an idea of the challenge these males face. Because the female spider is so much larger, her genital opening is consequently large as well. So to be able to mate successfully with a female, this little male spider has to possess genitalia large enough to match her opening.

Spider genitalia are perhaps not what you might expect. They are palps —appendages akin to modified limbs—that place a packet of sperm, the *spermatophore*, inside the female's genital tract. Because these palps have to be disproportionately large relative to the male to mechanically couple with the female, each palp accounts for about 10 percent of the male's body mass. There are also two of them. This is a problem for the male cobweb spider because before he can have sex with a female, he needs to find her. This means traveling often large distances (for a small spider) while hobbled by enormously heavy genitals that not only slow him down but soon wear him out.

Male *Tidarren* spiders solve this problem in the most pragmatic way possible: they rip one of their own oversized palps off. They do this shortly before they reach full maturity and just after molting off their old, too-small

2. Based on some of the more adult parts of the Internet, some of you already have.

exoskeleton. While the new exoskeleton is still soft and hardening, the male secures one of his two palps to a silk scaffold and then turns in circles while pushing at the bulb of the palp with his third and fourth pairs of legs. In this way he twists off that palp, with the twisting motion serving a secondary function of closing the wound, preventing excessive fluid loss and infection.

The results of this extraordinary self-amputation are no less extraordinary—male sprint speed increases by 44 percent following palp removal, and endurance increases by 63 percent! Male spiders who have removed a palp can run faster and for longer than those who have not, and this gives them a clear advantage in terms of eventually locating a female to mate with. Most strikingly, however, the distance they can run before they become exhausted is increased by roughly 300 percent compared to males with both palps. This is enormous. A movement distance of just 2 m (6.5 ft.) for a little male *Tidarren* spider works out to be roughly 1,400 body lengths, or the equivalent of a person moving 2.5 km (1.6 mi.), so tripling that distance is no small feat. Indeed, the consequences of exhaustion for these animals are also huge, and males with two palps die more frequently following exhaustion than males in the one-palp condition. As gruesome as it may seem, palp removal has substantial benefits in terms of increased movement speed and distance, mate location and competition, and even survival.

The Burden of Motherhood

The negative influence of extra mass on burst performance is pervasive, as we have seen. However, males with genitals large and heavy enough to impair locomotor performance are uncommon, regardless of what men on dating websites would have women believe, and the effects of mass on locomotion are really more of a problem for pregnant females.

Pregnancy involves female animals carrying around little developing animals inside them, and the costs of doing so are not trivial. It has not escaped the attention of biologists that performance abilities ranging from running to swimming are often impaired in pregnant females. Female zebra finches who lay large clutches of eggs, for example, take off more slowly when pregnant compared with females who lay smaller clutches of eggs. However, although it might seem obvious and intuitive that this decreased take-off speed is due solely to the added weight of the eggs, there are other reasons why gravid (pregnant) females might suffer performance costs. For instance, pregnancy brings about a host of other physiological and hormonal changes, any number of which might have equally or even more important effects on performance than offspring or egg mass. In the zebra finches, compromised performance persists even after the eggs have been laid and is accompanied by changes in flight muscle volume. Specifically, the flight muscles of females who lay large egg clutches shrink compared with those of females who lay smaller clutches, which accounts for the decreased take-off speed after laying. This is an example of a *life history trade-off* (more on this in Chapter 8), and it happens because resources that would usually be used to support the mother's flight muscle function are diverted and invested in her eggs instead.

An elegant experiment in a very different animal assessed the relative importance of clutch weight versus other factors. Researchers at the University of Sydney found that injecting sterile fluid into the peritoneal cavity (the space in the abdomen where eggs would be stored) equivalent to 25 percent of body mass (roughly the mass of a clutch of eggs) in Australian garden skinks clearly decreased the sprint speeds of the treated lizards independent of the physiological milieu that accompanies actual pregnancy, thus confirming that the physical burden of pregnancy alone can clearly influence performance even without those additional (but probably still important) physiological factors.

Because the locomotor demands placed on females during gestation by their increased mass are not trivial, in some cases there appears to have been selection to compensate for them. One such case was described by Jeff Scales, now of California State University, Stanislaus, as part of his dissertation research. The clutches of female green iguanas can range from 31 percent to 63 percent of their nongravid body mass—a significant burden indeed, and one that should diminish the performance of female iguanas who do become pregnant. Surprisingly, this does not happen, and gravid female iguanas match the acceleration of nongravid females from a standstill.

Newton's second law tells us that acceleration is force divided by mass. This means that for a gravid iguana to match the acceleration of a nongravid female, she has to increase her limb muscle force production by the same amount as her mass has increased due to her eggs. This seems to be a tall order, because the limbs of nongravid females are already producing an impressive amount of mechanical power during acceleration. Work is done on an object when a force applied to that object moves it some distance, and power is the rate at which work happens. Scales found that nongravid female iguanas produce a peak total power of 667 watts/kg in each hindlimb (or, for any nonmetric types, about 0.4 horsepower/lb.). That's pretty good for an animal that weighs approximately twice as much as an NBA regulation basketball! But the peak total mechanical power output for pregnant iguana hindlimbs is very nearly double.

How does this happen? Accelerating from a standstill, as these iguanas are, requires a lot more power than steady-state locomotion (that is, locomotion at a constant speed), so it is hard to see how an individual could double its power output during acceleration. The most likely explanation is that the extra force required comes from the additional mass of the clutch itself, for two reasons.

First, external loading increases short-term power output because the additional mass stretches the muscles, and active stretching increases muscle force production. For any given animal there is therefore an optimal loading for power output, and if nongravid iguanas are suboptimally loaded for maximum mechanical power output, then the extra clutch mass would load the locomotor muscles in such a way as to prestretch them, enabling even higher force production and power over the duration of a stride. This argument is bolstered by the observation that male and female iguanas are shaped differently, with females having relatively shorter limbs than males, giving them a mechanical advantage in terms of this load bearing.

Second, iguanas overcome an inherent constraint in muscle function by swinging their legs faster. Another way to think of power is as the product of speed and force, and because of the way skeletal muscle works there is a fundamental trade-off between the speed and force of muscle contraction. You can contract a muscle very quickly or very forcefully, but you can't do both at the same time, and so the sweet spot for muscle power is to contract at intermediate speed and force with a particular contraction frequency. Gravid female iguanas that exert higher downward muscular force due to their extra mass therefore necessarily exert that higher force over a longer time, but they compensate for this by swinging their legs up to 21 percent faster than nongravid females. This allows them to apply higher forces for longer without increasing stride duration, giving them more time for force production over a given distance compared to nongravid females. All of this has to come at a price, though, and although pregnant females avoid a reduction in absolute performance as I have described, these adjustments should increase the energetic costs of locomotion significantly, which means that gravid females are probably more limited in the duration and number of sprinting bouts compared to nongravid females.

For most animals that are not green iguanas, however, performance tends to decline in pregnant females, forcing those females to adopt behavioral strategies for escaping predation that do not depend on locomotion. Female collared lizards change their behavior when gravid, staying closer to refugia when out and about, making it all the easier to hoof it into a protected area when threatened but also harder for those females to locate food sources. Pregnant female *Zootoca vivipara* lizards also appear to rely far more on crypsis than locomotor performance to avoid predators, disguising themselves rather than running away, as do pregnant female garter snakes.

The Shape of Sexual Conflict

The different costs of sexual reproduction paid by males and females lead inexorably to a line of thought that used to receive little explicit consideration: the reproductive interests of males and females seldom align.

On the face of it, this statement appears absurd. Isn't the whole point of sex for males and females to make babies together? That it is, but although it may appear to the casual observer that males and females are willing accomplices in reproductive acts that benefit them both, beneath this facile veneer roils a world of conflict. In many animal species sex is anything but a cooperative venture, and discord between males and females is the rule rather than the exception. This conflict arises via two main concerns: the identities of the specific males whom females would prefer to make babies with, and the expression of traits shared by males and females that each sex uses in very different ways. It is the second of these two issues that I will consider here, not because it is any more important than the first, but because we currently happen to know quite a lot more about this form of sexual conflict as it pertains to performance and athletic ability.

Despite the ubiquity of sexual dimorphism, males and females of a given species do nonetheless usually resemble each other and share many characteristics in common. This is because both males and females are built from a single *genome*, the set of genetic instructions that specifies how to assemble and grow a certain species of animal. This shared genome explains such evolutionary puzzles as why men have nipples despite having no use for them—women need nipples for breastfeeding, and since all human fetuses are potentially female before the SRY gene on the male's Y chromosome kicks in and causes the production of bucketloads of testosterone, it is far simpler for all fetuses, male and female, to possess nipples in the early stages of development than it is to turn those genes on in one sex but not the other. The reason this works is because having nipples doesn't come with any downsides for men. They don't do much, but there isn't anything about having nipples that is expensive or dangerous for men either, so they are not selected against. So the answer to the question "Why do men have nipples?" is simply, "Because women need them, and because men's having them is fine."

For certain other shared traits, though, this is not the case. In some animal species, the same structures in males and females do very different things, and thus selection acts differently on them in each sex. This does cause problems, because the same genes affecting that trait are being pulled in two directions—male-optimal and female-optimal—that might well be diametrically opposed to each other. The evolutionary tug-of-war that is set up in this way is one that neither sex can afford to lose. If selection favors the male version of the trait too strongly, then the female version of it is suboptimal, perhaps even maladaptive, and vice versa. This tension between the sex-specific needs of males and females is the driver of an important male-female genetic conflict called *intralocus sexual conflict,* and it is best illustrated by an example. Fortunately, some particularly good ones happen to involve animal performance, just in case you were worried that this book had suddenly turned into a tome entirely about evolutionary theory.

Wing length in birds is an important trait affecting aerodynamics and flight performance across a bevy of selective contexts. The great reed warbler is a migratory Eurasian bird with minimal sexual dimorphism apart from a tendency of the male to erect the feathers on his head when singing. Nonetheless, wing length in these birds is under opposite direction in males and females, with selection favoring longer wings in males and shorter wings in females. Selection on wing length is uncoupled from that for body size, meaning that it does not simply reflect an evolutionary advantage for males to be larger than females. Rather, males do well when they have long wings and females do better with shorter wings regardless of how large they are.

This opposing selection for wing length in males and females stems from the way each sex uses its flight capacity. Males and females differ in the timing of the spring migration, with males arriving at the breeding grounds up to two weeks before females. Arriving early is important to males, because the males that get to the breeding area and grab the prime territories first get to mate with more lady warblers than latecomers. Migration flights are more efficient for males with longer wings, creating positive selection on male wing length. Females, in contrast, do not race to get to the breeding grounds, but once they do arrive, they spend more time than males searching for food within dense reed habitat where maneuverability is paramount and long wings are a hindrance. Females require shorter wings to navigate this dense environment.

Reed warbler wings exemplify intralocus sexual conflict in that they are the same trait constructed from the same genes in males and females but have a different optimal size depending on which sex those genes are expressed in. Selection on males is currently stronger than that on females within wild populations of great reed warblers, which means that females are currently losing the genetic tug-of-war with males and are likely suffering reduced maneuverability as a result.

If intralocus sexual conflict results in suboptimal outcomes for one sex or the other, then should there not be selection to ameliorate these costs in the losing sex? Indeed there is, and sexual conflict can be resolved, or at least mitigated, in several ways. One way is through the evolution of compensation for those costs, as illustrated by several genera of incredible and fascinating little insects in the fly family Diopsidae, more commonly called stalk-eyed flies. This name, descriptive as it is, demands some explanation. Male stalk-eyed flies possess bizarrely shaped heads from which two long, thin stalks called peduncles protrude laterally on either side (fig. 4.2). The animal's eyes sit on the very end of these peduncles. That is, their eyes are literally on stalks sticking out from the sides of their heads! Although both male and female diopsids possess eyestalks, male eyestalks are much longer than those of the female in several sexually dimorphic species. In those species where males have larger eyestalks, they can be up to twice as large as stalks in the nondimorphic species.

The reason for this outlandish arrangement, as for most others, is sexual selection. These male eyestalks are actually ornaments, and females pay

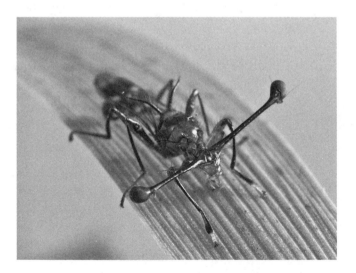

Fig. 4.2. A male *Teleopsis dalmanni* stalk-eyed fly.
Photo by Rob Knell.

close attention to the length of male eyestalks and prefer males with longer such stalks over males with smaller eyestalks.[3] Males also use their eyestalks as an assessment signal during male combat, resulting in some amusing eyestalk comparison behavior between rival males that usually works out in favor of the male with the longer eye span. But stalk-eyed flies also need to, well, fly, and just as the long tails of male widowbirds impair their flight performance, having two really long things sticking out of the sides of your head doesn't do you any favors either when you need to take to the air. Males of the dimorphic species *Cyrtodiopsis whitei*, for example, have a harder time taking off, showing both shallower angles of ascent and slower vertical velocity, compared to males from a monomorphic species in the same genus. Eyestalks also affect maneuverability; males in dimorphic species with longer eyestalks experience a large moment of inertia for roll and yaw, which means that they need to apply more torque to rotate their bodies and turn during flight than would be the case had they shorter eyestalks (or, even better, no eyestalks at all).

To deal with these performance issues, these males have evolved larger wings (in terms of both wing length and wing area) in tandem with their long eyestalks. Larger wings compensate for the flight costs imposed by eyestalks such that in practice they show no performance decrement. The relevance to intralocus sexual conflict comes when we consider this situation from the female perspective. In those very dimorphic species such as C. *whitei*, males need those long eyestalks to compete with other males and to attract the ladies, but the females themselves have little use for them—they experience similar flight costs as the males do but none of the benefits. However, they cannot simply divest themselves of eyestalks; breeding experiments have shown that if you raise stalk-eyes in the lab and select for longer eyestalks in males, females' eyestalks get longer, too, which suggests that eyestalks are

---

3. Although the eyestalks are also strangely functional. Most species of diopsids hang out on vertical plant stems or root hairs. When you go to collect them, their favorite escape strategy is to scuttle around to the other side so as to place the stem between themselves and you; then they look at you from around the stem with their stalk eyes while still remaining out of sight. It's charming the first couple of times that happens, but it gets old quickly.

indeed under similar genetic control in males and females. Eye span and wing length are positively related to each other in the females of several of these species, which means that they compensate as males do for their eye span being pushed beyond their biomechanical optimum by selection on male eye span. However, the lack of an increase in wing area in females relative to males implies that this compensation in females is incomplete. There is considerable variation among genera and species of stalk-eyed flies for compensation of intralocus sexual conflict on eye span; some compensate in the way I just described, whereas others do not, and among those that do there is evidence to suggest that there are multiple compensatory strategies. So the females of certain diopsid species pay the athletic price of their own preference for male eyestalks to a greater extent than others, depending on how they compensate.

Before leaving this topic, I want to talk about one other instance of this phenomenon that exemplifies the friction between survival and reproduction. Because in the vast majority of vertebrate animal species it is the females who lay eggs or give birth, the optimal shape of the pelvis is different for males and females. The problem here is intuitive; for a female to push the eggs or offspring out of her own body through whatever reproductive opening is used for that purpose in a given species, the eggs or offspring must pass through an opening in the pelvis. Animals vary in their mode of locomotion and the gaits that they use, but for many terrestrial animals it is generally best if the pelvis is narrow, because a wide pelvis results in splaying of the hindlimbs and a consequently inefficient locomotor gait— splayed limbs means that more force has to be exerted sideways by each limb as well as downward to keep the animal moving forward in a straight line. Narrow hips are good for males, then, but a problem for females because narrow hips leave a small pelvic opening through which to push eggs or small animals. Here is another situation where the optimal trait expression is different for males and females.

Females therefore face a trade-off: either a narrow pelvic aperture and efficient locomotion, but necessarily smaller offspring or eggs, or a wider pelvic aperture and large eggs or offspring, but compromised locomotion. This situation has received a lot of attention in humans in particular, where it is referred to as the *obstetric dilemma*. Compared to our ape relatives, women have a tough time in childbirth because of the large size of human baby heads, a consequence of long-term selection for large brains. Humans are born at a relatively early developmental stage because if we were born any later our heads would be too large to fit through the pelvic aperture at all. This is also why human babies are so useless compared to the infants of most other animal species, which, even at a very early age, can do other things besides sleep, cry, and vomit on you. The situation was probably exacerbated by our bipedal method of locomotion, the evolution of which necessitated extensive reorganization of our pelvic architecture that brought the contrasting morphological demands of parturition and locomotion into even greater conflict in women. So serious is the obstetric dilemma that it has been coopted as an adaptive explanation of male preferences for female body shapes. Males are attracted to females with wide hips, this controversial line of reasoning goes, because such women have an easier time in childbirth than narrow-hipped females. Of course, those women with hourglass figures do then bear functional costs in terms of a less efficient running gait. Or so it was once thought.

Despite evidence for this conflict working exactly as I have described in organisms such as lizards and tortoises, a study led by Anna Warrener of Harvard University argues that humans have overcome the functional consequences of the obstetric dilemma, and that although increased pelvis width in females does indeed affect the mechanics of locomotion, it does not result in hugely increased locomotor costs in females relative to males. Warrener and colleagues revisited the mechanics of the pelvis and found that the original model of the forces acting on the hips during locomotion in males and females was flawed. A more appropriate biomechanical analysis, which this time also involved measuring the forces during steady-state locomotion as opposed to inferring them, as had been done previously, shows that those forces are less disparate in men and women than previously thought.

There are some important caveats to this finding. First, these researchers were concerned with sex differences in locomotor efficiency, not in speed, and thus all individuals were measured walking and running at the same speed. Furthermore, it is unclear whether these results would hold outside laboratory conditions when individuals are faced with navigating inclines or while burdened. Nonetheless, it appears that the obstetric dilemma is much less of a dilemma for females in terms of locomotion than has traditionally been thought, although the jury is still out on whether this is because females compensate for these effects, as stalk-eyed flies and female green iguanas do, or because the effects of pelvic width on human locomotion were illusory all along.

Juicing Up Performance

Many of the differences between males and females in size, shape, and behavior are rooted in physiological differences that are either directly or indirectly related to the different reproductive roles of each sex. Sex differences in physiology have received surprisingly little attention and can be subtle. Male alligators, for example, have more active mitochondria—the little cellular components that use oxygen to provide cells with energy—than female alligators during the breeding season, possibly to fuel the higher seasonal demand for locomotor performance in males. But perhaps the most powerful and obvious of these are the differences in production and maintenance of various hormones, of which the steroid hormones in general, and testosterone in particular, are by far the most important in vertebrate animals. They are also the best known in terms of performance, primarily because of their use and abuse by human athletes.

The original choice of testosterone supplements in various forms as performance enhancers was by no means capricious. Testosterone has several clear effects on animal characteristics that are germane to athletic performance. In male vertebrate animals, testosterone stimulates the development of primary and secondary sexual characteristics through what are termed *organizational* and *activational effects*. Organizational effects prompt permanent changes that are wrought early on in development and within a very specific developmental window, whereas activational effects are temporary and occur in adults. Organizational effects of testosterone underlie many of the widespread sex differences that we see in the natural world; that males are often larger, more muscular, and more aggressive than females is accounted for in no small part by higher levels of testosterone in males of many species during development. However, activational effects are just as important, and it is not uncommon for males to ramp up testosterone production at certain times. The males of many species of birds and lizards, for example, exhibit a spike in testosterone production in early spring, just before their breeding season, as would benefit animals that will be spending the coming months competing over access to females.

Among the best-known activational effects of testosterone is the ability of this hormone to spur skeletal muscle growth, which has obvious implications for performance. This phenomenon is most understood in humans, but the detailed cellular mechanisms by which it operates are equally poorly known in all animals. What we do know is that testosterone causes muscle cells to increase protein synthesis, resulting in muscle fiber growth and, ultimately, an increase in muscle mass. Testosterone also causes a dose-dependent increase in the cross-sectional area of muscle fibers, an important functional characteristic of muscle that is directly proportional to its capacity for force production (see Chapter 6 for more detail). These changes at the cellular level translate into clear effects on performance, with studies on humans reporting improvements in strength ranging from 5 percent to 20 percent with testosterone supplementation. So steroid hormones do indeed cause individuals to become stronger and more muscular, both of which are desirable traits for athletes competing in many sports, and this is why steroid supplementation is banned in most professional athletic leagues.

These dramatic effects of testosterone on muscle development and, ultimately, performance are also why women have a much harder time building muscle in the gym than men do—with significantly lower testosterone production and consequent lower levels of circulating testosterone in women compared with men, gains in muscle mass and strength in women are earned at a higher price in terms of time and effort than in men. This has led to repeated suggestions that women have more to gain than men do from steroid supplementation. Even in men, however, testosterone-driven muscular gains are not free, and the greatest increases in muscle mass and performance are seen in individuals who combine steroid supplementation with a resistance-based exercise regimen. Indeed, strength gains with exercise alone are roughly comparable to those with testosterone supplementation alone—still an impressive effect.

A final important lesson from studies of steroid use in humans is that although steroids have marked effects on strength, they have no impact on endurance performance. This is consistent with the general physiological principles underlying burst performance versus endurance: strength-based performance traits are dependent on force production and, therefore, muscle, whereas endurance capacities are more heavily dependent on oxygen delivery. So if an athlete wanted to artificially increase performance in an endurance event, she or he wouldn't necessarily receive any benefit from steroid supplementation but would instead consider methods that enhance the quantity and longevity of oxygen-carrying red blood cells or otherwise increase oxygen delivery capacity. Endurance training at high altitudes or in artificial environments that mimic the low partial pressures of oxygen at altitude accomplishes this naturally through acclimation, resulting in increased aerobic capacity through physiological changes including growth of the heart and elevation of the red blood cell count. But similar effects are achieved by supplementing with EPO (erythropoietin), a hormone that increases red blood cell production; introduction of HIF (hypoxia-inducible factor) stabilizers that increase vascular growth; or the removal and storage of red blood cells for injection back into the bloodstream before an event, called blood doping.

Testosterone supplementation has clear effects on performance in nonhuman animals, just as it does in humans, and leads to increases in muscle mass in vertebrate animals as varied as fish, frogs, birds, small mammals, and lizards. Of course, animals such as frogs and birds are not bulking up with testosterone creams and injections on the sly (that we know of), but hormone levels do nonetheless vary seasonally in many species. We can understand these seasonal changes and clarify the hormonal effects on different traits by manipulating hormone levels through experiments. The effects of steroid supplementation seen in humans are largely mirrored in nonhuman vertebrate species, but a wider, comparative perspective also reveals enormous variation among animal species in testosterone levels and function. For instance, testosterone supplementation increases both sprint speed and endurance in male *Sceloporus undulatus* lizards, counter to the situation in humans. In male brown anoles, however, testosterone affects neither endurance nor sprint speed but instead has positive effects on bite force, which as we have seen is an important trait for male lizards in terms of male combat. In the lizard *Gallotia gallotia*, by contrast, testosterone supplementation increases general muscle mass but does not affect sprint speed or bite force. (Testosterone-supplemented males did grow larger penises, for whatever that's worth.)

This selectivity of testosterone in diverse species means that differences between males and females in performance might be accounted for by hormonal effects on some traits but not others, and that testosterone is not a catch-all explanation for such differences. In some cases, superior performance in one sex compared to the other might be better accounted for by shape, size, or even motivation (although mounting evidence suggests that motivation itself is under partial hormonal control). Furthermore, supplementation studies are often performed in laboratory settings where the animals aren't doing much, which makes it tricky to detect subtle effects.

Despite this interspecific variation, it is clear that organization and activational differences in testosterone do account for sex differences in certain performance traits and that these differences occur in tandem with sex-specific testosterone levels. But even here we see wide variation among animal species. In some species of socially monogamous birds, for example, females exhibit absolutely higher levels of testosterone than males because of the need for these females to compete over males and breeding sites instead of the other way around. Little is known about the effects of artificially increasing testosterone on female performance,[4] but testosterone supplementation does increase both body mass and skeletal muscle mass in some birds. There is also evidence that testosterone levels in females change in response to selection for increased circulating testosterone in males. Just

4. The documented doping of female athletes by the East German regime between 1970 and 1986, which won the nation Olympic medals all the while wreaking havoc on the athletes' health and personal lives, was in no way proper science.

like many other cases of intralocus sexual conflict, such as wing length in warblers, increasing testosterone expression in males can have negative consequences for the females because testosterone has many other effects besides those on muscle mass and performance. For instance, testosterone is a notorious suppressor of immune function, often leading to increased levels of parasitism, and over the long term it reduces growth, impairs reproduction, and depletes energy reserves. These negative effects are manifest in the females of species where males are selected for high testosterone, such as zebra finches, where unresolved sexual conflict over high testosterone in males causes high testosterone in females, which in turn inhibits female growth and reduces fecundity.

These various trait and sex-specific effects of testosterone highlight the complexity of hormonal effects within living organisms. Certainly, testosterone is not the only hormone that affects animal locomotor performance. To give but one example, treatment with corticosterone, a hormone released in response to stress, increases stamina in mammals, birds, lizards, and turtles, probably through mobilizing stored carbohydrate and fat reserves to fuel endurance performance. Corticosterone also inhibits testosterone in some species. The effects of various hormones both on each other and on the traits they influence may therefore be contrasting, multiplicative, or neutral depending on the animal in which they are expressed.

Effects of a given hormone are thus often highly interactive and dependent on the overall hormonal milieu. Since this milieu is frequently different in males and females, again because of the different reproductive roles in each sex, hormones have complex, but nonetheless powerful, sex-specific effects on both performance and a host of related traits, all of which contribute to the diversity of sex differences in the natural world.

FIVE

Hot and Cold

The premise of this book is that natural and sexual selection shape animal athletic abilities to successfully navigate both environmental and social obstacles, and thus far I have illustrated this with several cunningly chosen examples and contexts. But performance is also deeply rooted in the physiological capabilities (and limitations) of the animal in question. Although some of these capabilities are based on sex, others are even more fundamental. For those of us who concern ourselves with performance in nonhuman animals, one of the most important factors affecting animal performance abilities is the physiological, and thus athletic, response to variation in temperature.

Temperature influences many aspects of biology, from the function of cells, tissues, and organs to the distributions of entire species. As humans, our awareness of thermal ecology is limited to certain specific contexts—whether we need to take a coat when we leave the house, what setting to put the oven on for a recipe, and who used all the hot water. In particular, unless we live in some part of the world that experiences extreme weather conditions (generally, either at high latitudes or near the equator) or until we become sick, we seldom think explicitly about our own body temperatures except in cases like visiting New Orleans in July, Minneapolis at any time that isn't July, or Antarctica any time at all.

I do not mean to imply that humans are in any way indifferent to temperature. One of the reasons for our current widespread geographic distribution across the globe is our ability to modify the immediate thermal environment to fit our purposes. But although variation in local environmental temperature causes us occasional discomfort, it seldom affects our daily activities or physiological ability to do certain things except under extreme conditions. For many other animals that lack the ability to construct air conditioners and heaters, awareness of variation in temperature is paramount, and failure to respond to that variation has serious or deadly consequences.

Why Butterflies Shake Their Wings and Lizards Bask on Snow

If you were to visit the forests of a tropical locale such as Costa Rica, you would doubtless notice the stunning diversity of butterflies. If you were then for some reason to get up early when the day is still cool and wander about in the forests (particularly at higher elevations) you would also notice these insects doing something remarkable—they sit on tree trunks or leaves and shake their wings rapidly.

You might think at first that this is some kind of display or an epidemic of lepidopteran Parkinson's disease. In fact, it is neither, and what these insects are really doing is shivering, in the same way and for the same reason that you or I might shiver, which is to generate heat via repeated muscular contraction, thus warming themselves up. But whereas humans shiver when we feel a chill, being a bit cold doesn't necessarily impair our performance abilities even if it does make us uncomfortable. We can still run, jump, climb, or play sports in all but the most frigid of temperatures, and some special people apparently even find it pleasurable to begin their day by voluntarily plunging themselves into freezing-cold water. For small insects such as butterflies, however, the predawn chill has a number of more serious outcomes, the most important of which for our purposes is this: a cold butterfly cannot fly.

To understand why butterflies must shake their wings to warm themselves, we first need to learn something about the relation between temperature and how muscles and other tissues that make up an animal's body work. This requires a brief detour from our main path exploring the sunny meadows of performance into the ironically more darkly wooded areas of thermal physiology. I promise it will be largely painless, though, and the insights we can gain into the effects of temperature on animal performance along the way are profound.

The mechanisms underlying and enabling tissue and organ function—be they muscles or kidneys or liver tissue or whatever—at the most basic level are chemical ones involving molecules interacting with each other. Like all chemical reactions, the rate at which these physiological reactions occur is affected by temperature. Living organisms possess a variety of specialized proteins called enzymes that are involved in a mind-bogglingly large array of physiological reactions. The job of enzymes is to facilitate those reactions, allowing them to proceed faster than they would if the enzymes were absent. A common feature of enzymes is that they are extremely sensitive to temperature, and they thus function optimally only in a certain range of temperatures. Beyond that range, enzymes operate poorly, slowing the rate of the reactions they normally expedite.

It is therefore a priority for organisms to maintain those enzymes within their thermal zone of optimal efficiency, and the best way to do that is to regulate the core body temperature (T_b) of the animal in question to ensure that it never becomes too high or too low. Animals such as our early rising butterflies that do allow their T_bs to stray too far from the optimal range overnight find that the reactions that power their flight muscles (among others) are occurring too slowly for those muscles to function properly —hence the preflight warm-up shivering. The efficacy of the preflight warm-up is impressive, and some butterflies warm up rapidly, compensating for a 23°C (73°F) deficit in around six minutes. Honeybees can warm up even faster, and will use this ability as a defensive strategy to heat-murder giant hornets that enter their hives. They do this by clustering around invaders and shivering their flight muscles to form a heat ball, causing the target hornets within to overheat and expire.

Okay, so butterflies have to be warm, usually at least 28°C (82°F) and preferably 33°C–38°C (91°F–100°F), for their muscles to operate efficiently enough for performance. (As always there are exceptions, and some small insects fly at extremely low T_bs, albeit not very well.) But so do humans, and although humans also shiver, we seldom find ourselves in a position where our muscles are too cold to function at all unless we are very, very unlucky. What is different, then, in this regard between humans and butterflies?

The short answer is that butterflies produce their own heat only intermittently when their muscles are active, whereas we humans produce large amounts of our own internal body heat all the time, shivering notwith-standing. Generating so much heat is an extremely useful feature of our physiology that places us, along with all other mammals and birds, within a category of animals called *endotherms*. Endotherms produce heat by har-nessing the inefficiency inherent in all biological systems. If a given process is 100 percent efficient, then that means that all of the energy that is put into that process is ultimately converted to work and none is wasted. In reality, total efficiency is elusive at best and energy is almost always lost—usually as heat. This is as true of biological systems as of any other, and the reason that shivering works to warm up an animal is because heat energy is produced in overcoming friction during movement and compensating for elasticity in muscles and tendons during those rapid contractions. Consequently, of the chemical energy that is put into a physiological process such as contracting a muscle, only some percentage of it (about 25 percent on average) is trans-formed into work. The rest of that energy is manifest as heat.

Endotherms deliberately run several "wasteful" physiological processes (including, but not limited to, shivering) that collectively serve the ultimate purpose of producing heat to warm their bodies. Many of these processes run not only under cold environmental conditions but constantly, incurring considerable energetic expense. Thus, endotherms are also characterized by having high resting *metabolic rates*, which is another way of saying that they use a great deal of energy every day just running their basic physiologi-cal processes, in exactly the same way that a car whose engine is set to idle at high revs uses more fuel than a car whose engine idles at lower revs. This heat production of endotherms explains why small desert-dwelling mam-mals do not greet daybreak after a chilly night by lying on their sides and twitching until such time as they are warm enough to stand up and move around. In fact, this feature grants most endotherms a degree of indepen-dence from environmental temperatures and allows them, in conjunction with specialized adaptations that limit heat loss from their bodies, to inhabit even extremely cold environments thanks to their ability to immediately replace lost body heat.

Producing one's own internal heat is clearly a very useful (albeit expen-sive) strategy. But as the butterflies show, it isn't the only one. Other ani-mals—in practical terms, pretty much anything that isn't a mammal or a bird, although there is room for disagreement here—forgo producing heat themselves in favor of the energetically cheaper approach of warming up their bodies using heat absorbed from the environment. These animals, called *ectotherms*, have a much more complex relationship with environmental temperature.

Because ectotherms are required to regulate their T_bs almost exclusively through heat exchange with the external environment, those T_bs depend greatly on the animals' behavior, including where, how, and for how long they position themselves in their local habitat. For example, many ectotherms adopt specialized postures to absorb heat from the environment, a behavior known as basking. Lizards in particular often shuttle continuously between cool and warm areas as required to maintain their T_b of choice, and marine iguanas famously alternate between warming up by basking on hot rocks and cooling off by diving into the ocean.

Basking and shuttling are among the battery of behaviors collectively referred to as *behavioral thermoregulation* that ectotherms use to adjust their body temperatures as required, and it is so effective that it allows some ectotherms to survive and perform even in some surprising places. To give but one example, the lizard *Liolaemus signifer*, which lives in the Andes at extremely high elevations (about 4,000 m, or 13,000 ft.) and thus cold temperatures, thermoregulates well enough to support locomotion by basking on mats of vegetation that insulate it from the cold ground beneath, and absorbing both the heat from the sun and the sunlight that is reflected off the snow. I want to say that again to be clear—these lizards warm themselves up by *basking on snow*, which still amazes me whenever I think of it. By contrast, endotherms rely less on behavioral thermoregulation and instead regulate their T_bs by means of circulatory and other adjustments collectively called *physiological thermoregulation*. These limit heat loss and gain, and allow endotherms to tweak their T_bs with remarkable precision to maintain them at a relatively constant value and within a very narrow range (although that range may differ depending on whether the endotherm in question is a placental mammal, a marsupial, or a bird). To be clear, though, the idea that endotherms thermoregulate physiologically and ectotherms do so behaviorally is another generalization, and there are many exceptions. For example, some ectotherms such as green turtles and crocodiles exhibit cardiovascular adaptations that allow them to adjust their circulation and physiologically thermoregulate in ways that endotherms cannot. Certain endotherm marsupials rely on basking to warm up at the start of each day, and Japanese macaques bathe in hot springs to stave off the chill of winter. (When I lived near Sydney's eastern beaches, the summer invariably heralded an influx of pale British tourists whose basking behavior on Bondi Beach was strikingly similar to that of the Galápagos iguanas.)

Although most animals fall squarely within either the ectotherm or endotherm category, intermediate forms of endothermy do exist whereby animals generate their own heat either only at certain times or only in certain parts of their body (a strategy called *heterothermy*). Some researchers might consequently take issue with my characterization here of butterflies as ectotherms, arguing that intermittent internal heat production via muscle shivering is a form of endothermy. I don't disagree, but I would argue in return that the primary mechanism of heat production is distinct from that seen in true endotherms such as mammals and birds, and this is probably the point where those of you who don't do this kind of thing for a living have stopped caring entirely. With that final caveat, let's get back to talking about performance.

Real Ectotherms Have Curves

The significance of maintaining an appropriate T_b compels ectotherms such as snow-basking lizards to go to some pretty unusual lengths to warm themselves in cool environments. Variation in environmental temperature is extremely important to ectotherms because, although they are adept at controlling their T_bs through altering their behavior, ectotherms are ultimately limited by the amount of heat available in their environment. Consequently, if the environment is too cold, the animal itself becomes cold, which slows down a variety of physiological processes, from its development and growth to its ability to digest food. (In some reptiles, the temperature at which the eggs are incubated even influences the sex of the embryo.) On the other hand, if the animal becomes too hot, the physiological machinery underlying those processes is damaged and ceases to function correctly. Of the processes most obviously and dramatically affected are those that promote athletic performance.

Temperature effects on ectotherm whole-organism performance follow the same general pattern as for other physiological traits described above and are encapsulated in the *thermal performance curve* (TPC). This curve can be drawn as a simple graph describing the relationship between the performance ability of interest and T_b. Figure 5.1 shows a generalized thermal performance curve for sprint speed in an ectotherm (clearly one that is able to sprint, but beyond that the specifics are unimportant). When this animal's T_b is very low—say 15°C (59°F)—that animal can achieve only about 40 percent of its maximum sprint speed. However, as the ectotherm becomes warmer, it is able to run ever faster, until it becomes warm enough to achieve its top speeds. Figure 5.1 also shows that there is a narrow range of T_bs over which at least 95 percent of this animal's maximum sprint speed

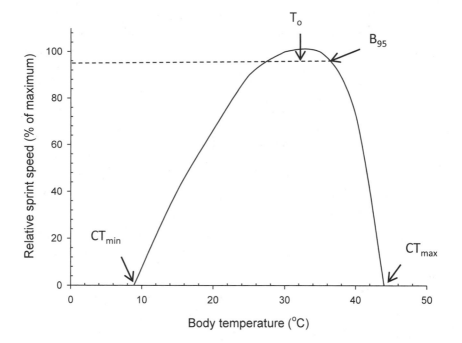

Fig. 5.1. A generalized thermal performance curve. CT_{min} and CT_{max} are where relative speed = 0, which corresponds to T_bs where the animal loses the ability to move. Optimal sprinting temperature (T_o) is the midpoint of the range (indicated by B_{95}) over which at least 95 percent of maximum speed is possible.

can be expressed; by convention, the midpoint of this range is known as the optimal temperature (T_o) for that performance trait. However, if body temperature increases beyond T_o, performance drops off very steeply.

The thermal performance curve is bounded at the upper and lower extremes by the temperatures where animals become either so warm or so cold that their muscles stop functioning and performance (indeed, any movement at all) is no longer possible. These are called the *critical thermal maximum* (CT_{max}) and *critical thermal minimum* (CT_{min}), respectively. A further characteristic of the thermal performance curve is that it is asymmetrical, with peak performance occurring at temperatures that are closer to the CT_{max} than to the CT_{min}. Although the shape of the thermal performance curve is similar across species, natural selection has modified the thermal sensitivity of many ectotherms. The details of the curve, such as the values of CT_{min}, CT_{max}, and T_o, are therefore species-specific and must be measured for each animal under consideration. For example, the tropical cichlid fish *Astronotus ocellatus* loses its ability to move below 13.6°C, whereas Antarctic *Trematomus* fish cannot even survive above 6°C (43°F).

In August 2016, reports began to appear in the media of cockroaches flying in New York City. It is no coincidence that these cockroaches began taking to the air during a heatwave. I haven't been able to turn up any studies that have measured the TPC for cockroach flight; however, given that flying cockroaches are an annual feature of life in the hot and humid American South (where they are known, possibly out of some deep sense of denial, as Palmetto bugs), one possible explanation is that temperatures and humidity levels New York City in August 2016 were high enough to allow cockroach flight in the area for the first time (or, more likely, to prompt them to fly frequently enough to be noticed doing so). The humidity could be especially important, because studies have shown that cockroaches are sensitive to desiccation and that humidity levels can affect their thermal preferences. If this is the case, then New Yorkers will have to learn to embrace their new flying cockroach overlords as temperatures continue to rise as a result of climate change.

The thermal performance curve has profound implications for fitness-related tasks involving performance. Consider how the performance traits I have described previously influence the outcome of fitness-related events, be they escape from a predator or combat between two males; now consider how those outcomes might differ if the animals in question are forced to perform at T_bs that do not allow for sufficient levels of performance to enable success.

The mating system of the freshwater eastern mosquitofish is dominated by sexual coercion, whereby males have foregone notions of gentlemanly conduct and achieve all matings forcibly and with no regard for the mating preferences of the females themselves. Female mosquitofish, however, are no passive victims, and behave aggressively toward inappropriately amorous males. Since these are fish, both male coercive abilities and female resistance to coercion are enabled by swimming performance, which is in turn affected by T_b. This suggests that altering T_b should also affect these mating behaviors in males and females, respectively, and work on mosquitofish by the Performance Lab at the University of Queensland shows that this is indeed the case: male success at coercing females to mate and the females' chances of resisting that coercion are both greatly enhanced at 32°C (90°F) as compared to 12°C (54°F). If one were to imagine a scenario whereby a warm male encounters a cold female, that male would have a much easier time of forcing himself on that female than if she were warmer and thus more capable of defending herself. Although this scenario is unlikely in these fish, which are small and experience little variation in the thermal environment, this is exactly what happens in some garter snake species at high latitudes. In these animals, cold females emerging from a long winter sleep find themselves mobbed by warmer males that have been active for longer, with ample opportunity to warm themselves. That same performance limitation leads some species of lizards to near a potential refuge when they are cool in case of sudden predator encounters, which means that not only can they run but they can hide as well.

Midnight Run

When ectotherms behaviorally thermoregulate, they do so to reach some target T_b. Based on the thermal performance curve, one might predict that this target core body temperature should be at or close to optimal temperature for one or more key performance traits, because that allows for maximum athletic performance. However, the T_bs that ectotherms prefer to maintain (which we can call T_{pref}) are usually slightly lower than T_o. Indeed, the flexibility of ectotherm physiology means that ectotherms can often choose to raise their T_bs to within that optimal range or not (environmental temperature permitting) and there are many valid reasons for not doing so.

Preferred and optimal T_bs do not perfectly coincide at least in part due to energetic considerations; an ectotherm having a T_b at the optimal temperature for sprinting has clear benefits in terms of achieving high sprint speeds, but it also costs more energy than having a lower T_b because warmer ectotherms spend energy at a higher rate than cooler ones. Spending energy profligately would be fine if these animals faced no energetic limitations and lived amid a perpetual buffet of food resources, but in nature this is seldom the case. Yet another reason that T_{pref} is often lower than T_o is the asymmetric shape of the thermal performance curve. Ectotherms are not perfect thermoregulators and cannot always attain T_o with precision; however, the consequences of maintaining a T_b that is too high are more severe than those of maintaining a T_b that is lower than optimal. Being too cold merely compromises performance, but being too hot and approaching CT_{max} risks serious physiological damage and possibly even heat death.

Yet another consideration is that behavioral thermoregulation incurs several costs. These include time or opportunity costs that come about because any time spent on the behavioral thermoregulatory chores of shuttling and basking is time that could otherwise be spent on other fitness-related tasks such as foraging or mating. Ectotherms that are active at night, however, face a different problem, in that they might never be able to reach their optimal temperature at all.

For ectotherms that are dependent on environmental temperature, it is sensible to be active during the daytime when it is warm. Most lizards, for example, are diurnal, which allows them access to a range of possible T_bs during the times at which they are most active, even if they live in the snowy Andes. A group of lizards called geckos, however, buck this trend. Almost all geckos are nocturnal and thus are active when temperatures are low and variable. The frog-eyed gecko, for example, is active at an average T_b

of 15.3°C (60°F)—very much at the lower end of its thermal performance curve, which allows it to achieve only 25 percent of its maximum endurance ability. (By comparison, the diurnal lizard *Platysaurus intermedius* is active at average T_bs of 27°C–30°C [80°F–86°F] during the summer, which places it at or close to its T_o, and within the ranges where it is capable of realizing at least 95 percent of its maximum capacity for both exertion and sprinting.)

Given this situation, one might predict that the shape of the thermal performance curve would have evolved to coincide with the range of environmental temperatures that are available to these animals. After all, if Antarctic fish are capable of dealing with the frigid environmental temperatures found in, you know, Antarctica,[1] then surely natural selection could have tweaked geckos to be able to achieve their maximum performance capacities at the lower T_bs at which they are most commonly active? In fact, this hasn't happened, and most geckos that have been tested have optimal temperatures for locomotion that are higher than they are ever able to attain in nocturnal environments, even with the aid of behavioral thermoregulation. This means in turn that geckos are almost never able to realize their true top speeds in nature during the times at which they are most active.

The significance of this constraint on gecko performance is not immediately clear; after all, geckos are very successful animals, having existed for more than eighty-five million years and constituting 25 percent of all described living lizard species, and although not all gecko species will likely be subject to these thermal constraints, many are. Despite the success of geckos as a group, their thermal physiology doesn't appear to have changed to complement their nocturnal lifestyle (although there is evidence that locomotion at low T_bs is cheaper for nocturnal geckos than for diurnal lizards at equivalent T_bs, so this isn't strictly true). Given that eighty-five million years seems like plenty of time for them to have adjusted their thermal physiology to run optimally at low T_bs, this suggests either that there are

1. The physiological mechanisms enabling this are too deep in the woods for our purposes here; suffice it to say that they appear to involve different versions of the same enzymes that work at higher temperatures, as well as a group of genes called heat shock genes.

some fundamental reasons why they have not done so[2] or, alternatively, that achieving maximum athletic performance isn't necessarily all that important to these animals.

Instead of altering basic aspects of their thermal physiology, nocturnal ectotherms could adopt another strategy of altering their behavior. Nocturnal insects, for example, are even more constrained than geckos by low nighttime temperatures because these insects tend to be small, and small things have very low *thermal inertia*. This is another way of saying that small things gain and lose heat much more rapidly than large things. As a consequence, the core body temperatures of very small animals can change a lot in a matter of seconds. This is why nocturnal moths are often covered with furlike scales: the fur serves as insulation that slows the rate of heat loss to the external environment, thus preserving the animal's flight abilities even at relatively cool nighttime temperatures. The combination of insulation, the aforementioned muscle shivering, and specialized heat exchangers that limit heat loss through certain parts of the body allow moths of the genus *Cucullia* in northeastern United States to maintain T_bs of 30°C–35°C (86°F–95°F) even in winter, when environmental temperatures are near 0°C (32°F)! But many other nocturnal insects lack these specialized adaptations and instead deal with cool nighttime temperatures more pragmatically by being active only in the earlier parts of the evening while it is still warm.

Why Lizards Can Bite You Even in the Cold

Whereas most performance traits to date are known to have a clear thermal performance curve, one such trait is notable for its independence of temperature effects. Within lizards, bite force is a key performance trait because it is used in a variety of disparate ecological scenarios. Lizard bite-force abilities can affect the type of prey they can consume, their ability to win or retain territories (as we have seen) and, importantly in some cases, to

2. One possibility raised by Ray Huey and others is that nocturnal geckos could still be exposed to high T_bs while sequestered in their daytime retreats, when they cannot thermoregulate. If CT_{max} and T_o are linked, then high T_os could be an incidental consequence of the need to tolerate high daytime temperatures.

intimidate and fend off predators. In fact, bite force is so important to lizards that it has evolved an exemption to the rule of thermal dependence, in that bite performance, unlike other types of lizard performance, is not affected by body temperature.

This observation has its roots in a 1982 study by Paul Hertz, Ray Huey, and Eviatar Nevo. Hertz and his colleagues showed that two species of agamid lizards flee rapidly from a perceived predation threat when they are warm but stand their ground and fight (which in lizards means bite) when their body temperatures are low. This behavior is likely an adaptation of lizards that live in open habitats and have little chance of hiding or outrunning a predator in the cold, leaving them no choice but to defend themselves. Some ectotherms that find themselves in this situation feign death instead, which works because many predators prefer fresh kills and are unlikely to consume already dead things. As an antipredator strategy, biting is effective indeed—in this study, Hertz ran the lizards while Huey wisely operated the computer, and Hertz paid the price of cool lizard aggression in terms of damaged hands from repeated painful agama bites!

But if these animals are cold enough that their locomotor abilities are curtailed, how is it they can still bite forcefully enough to deter a predator? The answer is revealed by a study on the agamid lizard *Trapelus pallida* led by Anthony Herrel of the Muséum national d'Histoire naturelle in Paris, which showed that bite force is not affected by body temperature in this species, whereas sprint speed shows the classic lizard pattern of thermal dependence, explaining nicely why these animals can rely on bite force to defend themselves at all T_bs. Intriguingly, this study also shows that the properties of the jaw adductor muscles responsible for generating bite force are somewhat different to those of the limb muscles powering locomotion, hinting at a physiological difference between jaw and limb muscle in these lizards that may ultimately give us insight into the functional basis of temperature dependence.

Dinosaurs!

With an understanding of some of the performance consequences of ecto-thermy under our belts, we are now ready to answer the questions that probably none of you were asking: Were dinosaurs ectotherms or endotherms, and how might their thermal physiology have affected their athletic abilities?

It is easy enough to look at a living animal and figure out whether it is an ectotherm or an endotherm, mostly because we know that if it is a mammal or a bird, it's an endotherm, whereas if it is anything else, it's an ectotherm (intermediate degrees of endothermy and heterothermy notwithstanding). But for extinct animals that we have never observed or measured alive, it gets trickier. The metabolic status of dinosaurs has been the subject of speculation almost since dinosaurs were discovered. The name dinosaur, as coined by the Victorian paleontologist Richard Owen, means "terrible lizard," and originally it was widely thought that dinosaurs were just like modern reptiles and thus that they had to have been ectotherms. The early popular re-creations of dinosaurs reflected this, portraying dinosaurs not merely as ectotherms but as explicitly cold-blooded, sluggish, and inactive animals (although Owen himself did not appear to subscribe to this view and consistently compared dinosaurs to mammals rather than lizards). This, coupled with a kind of temporal chauvinism that assumed dinosaurs were clearly inferior because they were (a) not mammals and (b) dead, resulted in absurd surmises such as the idea that *Brachiosaurus* lived its life wading in water because it was too weak to support its own enormous body weight. The realization that this was probably not the case at all was a long time coming, but inklings appeared very early on.

The first piece of evidence that dinosaurs, though in many cases terrible, were not actually lizards was the discovery in Germany in 1860 of a fossil that came to be known as *Archaeopteryx*. Although a fossil of the entire skeleton was eventually found, *Archaeopteryx* was originally described on the basis of a fossil of but a single body part: a feather. This was amazing enough, because feathers seldom fossilize. It seems that this particular feather had been fossilized in limestone, which is ideal for preserving more fragile tissues, hence the full scientific name of the animal, *Archaeopteryx lithographica*.[3] But even more amazing is what this finding implied, which

3. Other fossils of varying completion (twelve of the time of writing) were subsequently found, almost all in the same region of Bavaria that is thought to have once been a large lake, hence the limestone deposition. The species name *lithographica*

was that the roughly raven-sized *Archaeopteryx*, dated at around 145 million years old, was in fact an ancestor of all birds, ravens included. Indeed, although several earlier feathered proavian fossils have since been discovered, many of them in China, *Archaeopteryx* long enjoyed the distinction of being the earliest known bird.

The reason the discovery of the first bird was so significant was partly because of the timing, the first fossil having been found just two years after the publication of Darwin's *On the Origin of Species*, and partly because *Archaeopteryx* looked like the product of a dalliance between a pigeon and a *Velociraptor*, exhibiting a clear mix of avian and reptilian features (fig. 5.2). *Archaeopteryx* came replete with some very birdlike characters, most notably feathered wings, a feathery tail-fan, and a fused collarbone (or furcula), but it also showed some very reptilian features, such as claws on the wings, teeth, and a long, bony tail. This mix of characters very clearly placed it as an important transitional form between birds and dinosaurs. The feathers were especially suggestive and sparked two long-running, related, and to this day mostly unresolved arguments: first over whether a dinosaur that had birdlike characteristics such as feathery insulation and that was clearly the ancestor of a lineage of endotherms was itself an endotherm; and second over whether *Archaeopteryx* could fly.

The dispute over whether *Archaeopteryx* was an endotherm or an ectotherm might seem tangential to the more germane question (for our purposes) of whether this animal could fly. In fact, these issues are intertwined, and the argument over the flight capabilities of *Archaeopteryx* hinges largely on the nature of this creature's thermal physiology. Some features of *Archaeopteryx* are consistent with modern flighted birds, such as asymmetrical flight feathers that resist torsion, reduce drag, and may have been able to generate lift, as well as broad tail feathers similar to those found in modern birds. This indicates that *Archaeopteryx* could have been able to glide—but taking off from the ground is another matter.

Counting against powered flight in *Archaeopteryx* is the lack of an enlarged (keeled) sternum or breastbone, which acts as an attachment point

---

therefore relates to the fine-grained limestone used in printing. The generic name *Archaeopteryx* translates as "ancient wing."

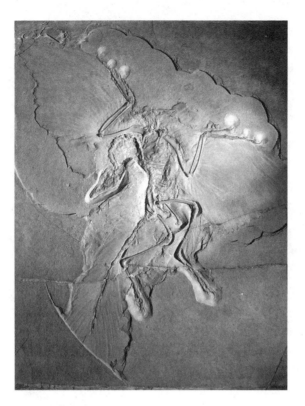

Fig. 5.2. *Archaeopteryx lithographica* fossil, currently in possession
of the Museum für Naturkunde, Berlin. The outline of the
wings is clearly visible in the limestone. Photo by H. Raab.

for the large pectoral muscles that power the downstroke of the wings in
modern birds. Another strike against flight is the particularly inconvenient
orientation of the shoulder joint (in combination with the lack of an important bird muscle called the supracoracoideus), which means that *Archaeopteryx* would likely have been unable to lift its wings up behind its back to
perform the upstroke part of flapping flight as in living birds. This led to the
so-called top-down hypothesis of flight evolution, which posited that proto-
birds such as *Archaeopteryx* must have used their clawed wings to clamber
up tree trunks and into treetops (as the juveniles of a modern Amazonian
bird called the hoatzin does) and glide from altitude (or leap and glide between treetops as modern gliding mammals do) because they weren't strong

enough to take off from the ground. It is also of note that bats do not have keeled sternums, and most cannot take off from the ground.

This is where the nature of the metabolic machinery becomes important. John Ruben of Oregon State University argued that despite lacking these avian flight adaptations, *Archaeopteryx* could well have flown, and possibly even taken off from the ground-up, if it had an ectothermic, reptilianlike physiology. Ruben's key insight is that because reptiles do not run those wasteful and energetically expensive processes that stoke the inner metabolic furnaces of endotherms and produce lots of heat, they use less energy at a given temperature than an equivalently sized mammal or bird. Because of that, reptile cells require fewer mitochondria, which you will recall are the little bits of cellular machinery that use oxygen to turn fuels like carbohydrates and fats into usable energy. And because they have far less mitochondria taking up space in their muscle cells than birds or mammals do,[4] reptiles can pack more of these nonaerobic muscle fibers into a given area than those animals can. This means that gram for gram, reptilian muscle is far stronger than endotherm muscle, as anyone who has ever handled a snake has quickly realized. So a reptilian *Archaeopteryx* might not have needed the large pectoral muscles of modern birds because its existing pectorals could have been plenty strong enough to power ground-up flight, at least for a short time.

A further possibility is that *Archaeopteryx* could have avoided the need for a powerful upstroke by adopting a downstroke-only flap-and-glide flight technique as used by living birds from swifts to red-footed boobies. Ken Dial and colleagues at the University of Montana have also shown that modern juvenile birds that cannot fly nonetheless use their feathered wings to generate downforce (in the same way that spoilers do on racing cars) when running up inclines. This affords them greater traction under these conditions, suggesting an initial locomotor function for wings that could well have led to the evolution of ground-up flight and that *Archaeopteryx* may have employed.

4. With some very specialized exceptions; for example, mitochondria make up only 3 percent of the volume of skeletal muscle in an iguana, and 2 percent of the volume of rattlesnake body muscle, but account for fully 26 percent of the volume density of the very rapidly contracting muscles in rattlesnake tails that shake the rattle up to one hundred times per second for relatively long periods.

Whether or not we accept that *Archaeopteryx* was an ectotherm, evidence from other sources supports the notion that this animal was capable of powered flight. For example, researchers have examined the degree of curvature in the claws on *Archaeopteryx* feet for comparison with the claws of modern birds that either perch on branches or spend most of their time on the ground, and found that *Archaeopteryx* claws are more similar to those of existing perching birds, implying an arboreal lifestyle. Analyses of the braincase and inner ear of the fossils also indicate that *Archaeopteryx* possessed characteristics very much in line with those of modern birds, such as an enlarged cerebellum (the part of the brain that deals with 3-D orientation, typically enlarged in animals such as birds and cetaceans) and an advanced inner ear for balance, further supporting the notion that it may have had at least rudimentary flight capabilities. Although many of these findings bolster the idea of flighted *Archaeopteryx*, the subject, as well as the ecto- versus endothermic nature of the beast, is still prone to the occasional heated debate and is by no means settled.

Hot and Cold Running Dinosaurs

Despite a great deal of argument, then, the idea was out there early on that *Archaeopteryx* might well have been an endotherm. And if the transitional stage between modern birds and dinosaurs was an endotherm, then why not dinosaurs themselves? This way of thinking really came into vogue in the 1970s and 1980s as a result of work by two paleontologists, John Ostrom and Robert Bakker. Ostrom and Bakker suggested that dinosaurs were not the slow, stupid behemoths they had been portrayed as but rather were active, warm-blooded, energetic animals—potentially even endotherms. These arguments were summarized in Bakker's 1986 book *The Dinosaur Heresies*. Controversial at the time, most of Bakker's arguments are now accepted by the scientific community, and the ideas Bakker put forward formed the basis of the movie *Jurassic Park* (for which Bakker himself was a scientific adviser). Since then, the notion that dinosaurs were endotherms has accumulated a great deal of evidence—some circumstantial at best.

Evidence for the endothermy of dinosaurs comes from several sources, all of which infer dinosaur physiology from fossils based on the known physiology of living ectotherms and endotherms. To give just a few examples, analysis of the types and ratios of different isotopes of oxygen in dinosaur bones suggests that tyrannosaurs likely had a constant T_b, a characteristic of most endotherms. Researchers have also estimated dinosaur growth rates from bone features and have shown that the growth rates of certain dinosaurs overlapped with those of modern endothermic animals. The feathered nature of many dinosaurs is also suggestive for reasons other than flight, because feathers likely evolved originally as a means of insulation and only became coopted by natural selection for flight later on. Animals with feathered insulation are unlikely to be ectotherms (especially if they are also small) because insulation is a double-edged sword; it increases the ability to retain metabolic heat, but it can also interfere with the animal's ability to absorb heat from the environment. This was brilliantly demonstrated in 1958 by Raymond Cowles of the University of California, Los Angeles, who, in a classic study (that I really wish I had thought of first), fitted out lizards with bespoke mink coats! Cowles found that while warmed furry lizards did indeed take longer to cool down than lizards that went au naturel, the fur coats impeded heat uptake in chilled lizards as expected. In other words, small, feathered, flightless dinosaurs would have had little to gain by insulating themselves with feathers unless they were also endotherms.

Although some of this evidence is persuasive, much of it is ambiguous. The oxygen isotope data, for example, is not a smoking gun for endothermy for two reasons: first, because some ectotherms also maintain relatively constant body temperatures, just as most endotherms do (a strategy called *homeothermy*), and second, because tyrannosaurs were very large. Just as small things have low thermal inertia and change temperature rapidly, so large things have high thermal inertia and change temperature extremely slowly. Indeed, for very large animals such as blue whales overheating is a real danger, and their biggest thermal challenge is losing heat, not gaining it, even in cold environments. This means that large tyrannosaurs could have maintained high and constant body temperatures by virtue of their size alone (a phenomenon called either *inertial homeothermy* or, more awesomely, *gigantothermy*). For instance, a 10-tonne (22,000 lb.) ectothermic dinosaur could have had a T_b of 31°C (88°F) in the same climate as an equivalent endotherm, even in winter.

However, important physiological differences still remain between ecto-therms and endotherms that are the same size and temperature due to the differences in the underlying metabolic machinery. Other analyses have made use of these differences to approach dinosaur endothermy from a different perspective: that of performance.

The mechanics of locomotion are well understood. Consequently, we can make decent predictions about the athletic abilities of animals from their shape and size alone, which is useful given that, in the case of dinosaurs, their shape and size is often all we know. This allows us to ask the question, "How athletic were dinosaurs?"

The late, great biomechanicist R. McNeill Alexander pondered just this question. He even wrote a book about it (the wonderful, though sadly out of print *Dynamics of Dinosaurs and Other Extinct Giants*). Thanks to Alexander and others who followed his lead, we now know, for example, that *Velociraptor*, with a top speed of 39 kph (24 mph), could definitely have out-run a human; that fighting *Stegoceras* may have exerted and withstood up to 9,000 N (0.9 tonnes) of force when butting heads during fights; and that pterosaurs were likely able to soar for considerable distances. As amazing as the performance abilities of extant animals are, it is even more amazing that these incredible animals once walked the earth and that we can divine their athletic abilities often based on little more than the shape and size of their bones and cunning applications of the laws of physics.[5] But our knowledge of dinosaur performance also allows us to turn our earlier question of dinosaur physiology on its head. In other words, instead of asking what dinosaur performance would have been, given a particular thermal physiology, we can instead ask, "Given inferred dinosaur athletic abilities, would these animals had to have been ectotherms or endotherms?"

The reason this is a valid question is, again, because ectothermy and endothermy allow for different physiological, and thus athletic, capacities. Most notably, ectotherms and endotherms differ in their capacity for aerobic,

5. But not always with certainty. For example, researchers have debated the running abilities of *Tyrannosaurus rex* for some time, and although some estimates peg the top speed of *T. rex* at about 28 kph (18 mph), others conclude that such speeds are impossible for an animal that may have been too large and heavy to run at all.

oxygen-supported performance. If we consider sprinting and marathon running, for example, we find that these two performance abilities not only differ in their speed and duration but are supported by different physiological pathways. Fast sprints are brief and happen so quickly that there isn't enough time to supply all of the muscles and organs involved with the oxygen required to convert stored fuels into energy. Instead, sprinting is enabled by a different physiological pathway that doesn't use oxygen at all but instead supplies that energy very rapidly, but in far smaller amounts. The trade-off here is that while that energy is quickly made available, high sprint speeds cannot be maintained for long. In human sprinters, most (about 79 percent) of the energy supporting the 100 m sprint comes from these *anaerobic* sources.

Slower, long-distance endurance running, however, is all about oxygen-fueled, aerobic locomotion, and consequently elite mammalian endurance runners have larger lungs, bigger hearts, and higher numbers of oxygen-carrying red blood cells—all things that enable high rates of oxygen supply to the muscles and other parts of the body that require it during those long, aerobically supported runs. One of the major constraints of an ectothermic physiology is very limited aerobic capacity, which probably has its roots at least partly in the fact (mentioned previously) that reptile muscle, for example, has less of the oxygen-using mitochondria that produce usable energy than endotherm muscle does. Different types of muscle fibers also exist, and reptiles have fewer oxygen-fueled muscle fibers, and more of the muscle type that is fueled by anaerobic sources (called glycolytic fibers). As a result, although reptiles are superlative sprinters, their oxygen-supported endurance capacities tend to be extremely poor.[6]

A 2009 study led by Herman Pontzer of Washington University in St. Louis took advantage of this limitation to estimate the energetic costs of running and walking at different speeds in thirteen bipedal dinosaurs (that is, those that run on two legs as opposed to four) ranging from very small to very large. He then asked whether those costs could have been borne by ecto-

6. The locomotor gait of lizards places yet another constraint on their aerobic performance. Lizards move with their limbs held out to the side, as opposed to directly underneath them like antelope. The resultant sprawling lizard running gait alternately compresses their lungs on either side with each stride, limiting their ability to breathe while in motion.

therms with limited aerobic capacities. Pontzer and his collaborators found that although small dinosaurs could have paid the energetic costs of aerobic locomotion had they been ectotherms, the aerobic capacities of ectothermic dinosaurs larger than 20 kg (44 lb.) would have been insufficient to meet their energetic demands for sustained locomotion. In fact, in the case of the very largest species included in the study, namely *Allosaurus* and *Tyrannosaurus*, those costs exceeded even the aerobic capacities of some modern mammals! This suggests that these larger species (at least) would have had to have been endotherms if they made extensive use of their aerobic locomotor abilities—something we can be far less certain of.

Several criticisms can be leveled against studies such as these. For example, because no ectotherms the size of *Allosaurus* or *Tyrannosaurus* exist anymore, these results rely on extrapolations of the aerobic capacities of small ectotherms to much larger sizes rather than on direct data, which is not ideal. One could also argue that modern animals that rely extensively on endurance have adaptations to reduce the energetic costs of locomotion, such as elastic energy storage (see Chapter 8), which dinosaurs may also have exhibited and are not accounted for. But given that we no longer have living, non-bird dinosaurs to measure, any study that tries to estimate the performance abilities of dinosaurs is necessarily forced to make similar assumptions and compromises, and so this criticism isn't entirely fair. In an attempt to circumvent at least some of these concerns, Roger Seymour at the University of Adelaide based a model of aerobic power generation on crocodiles, which are definitely not extinct; are also ectotherms; share a close evolutionary relationship with dinosaurs; and grow large enough to become relevant benchmarks for gigantothermy. Seymour's purpose was to test the oft-floated notion that having a high, stable T_b through gigantothermy bestows the same aerobic benefits as an equivalent T_b achieved through endothermy, but at a cheaper energetic cost. The results did not support this idea; Seymour found instead that gigantothermy results in ectothermic animals with much poorer endurance abilities than endotherms of the same size and T_b.

These two independent lines of evidence suggest that true endothermy is a key prerequisite for any large dinosaurs that relied on endurance in their day-to-day lives. These findings are also relevant to the much broader question of why endothermy evolved at all. The evolution of endothermy is yet another contentious area deserving of much more detailed treatment than I have room for here. However, even though we don't know exactly what selective factors drove the evolution of endothermy from ectothermic ancestors, we have a couple of ideas. The results of these studies, if correct, lend support to one of the leading hypotheses in particular, which is that endothermy evolved in response to natural selection for increased aerobic capacity. This boils down to the notion that true endothermy probably evolved to support the high levels of endurance-based locomotor performance currently seen in many birds and mammals—but that may have also arisen in dinosaurs.

We may never nail down the physiology of dinosaurs for certain without having actual dinosaurs to look at and measure. But the question "Were dinosaurs ectotherms or endotherms?" needs to be placed within proper context. "Dinosaurs" were not just one animal but a large and diverse lineage that existed over several geological time periods. Some were very small, others were very large, with all sizes in between; some were herbivores, some were carnivores, some were omnivores; some flew, most didn't; some were probably social, some solitary. We have already seen how thermal physiology can be considered to be a spectrum from ectothermy to endothermy, with inertial homeothermy (and several other categories that I haven't mentioned) in between.

With that in mind, consider this: dinosaurs existed for more than 230 million years and dominated the globe for around 135 million of those. These enormous time periods are difficult for us to wrap our heads around, and indeed a common assumption is that most of the famous dinosaurs were contemporaries. In fact, there is more time separating *Tyrannosaurus* from *Stegosaurus* than there is between *Tyrannosaurus* and us, which means that the awesome heavyweight dinosaur fight between *Stegosaurus* and *T. rex* that many of us imagined as kids never occurred.[7] One of the lessons from evolutionary biology is the tremendous amount of diversity that can evolve in animal lineages, sometimes within remarkably short periods of time. The

7. Science: ruining your childhood since the seventeenth century.

number of dinosaur species that actually existed is unclear, but about three hundred dinosaur genera are currently known, and there is reason to believe that seven hundred to nine hundred more genera remain to be discovered.

Given this massive diversity, it is by no means inconceivable that dinosaurs could have evolved a range of thermoregulatory strategies, including true endothermy, multiple times in different lineages over at least the 135 million years that they were the undisputed masters of the animal world. Viewed in this light, the answer to the question "Were dinosaurs ectotherms or endotherms?" is probably "Yes" or, if you prefer more nuance, "It depends."

Heat Storage and Locomotion

So far, I have talked exclusively about how temperature affects the performance of ectotherms. My neglect of endotherm T_b effects is because endotherms don't show thermal performance curves for locomotion, which renders their athletic abilities less interesting from a thermal perspective. But endotherms, and large mammals in particular, can suffer severe consequences of increased T_b during locomotion in certain circumstances.

Just as muscles produce heat during shivering, so the repeated muscular activity associated with locomotion also warms the animal. The extent to which this happens depends on the size of the organism in question, as well as on the type, intensity, and duration of activity. Because large mammals are prone to losing that activity heat much more slowly than smaller ones (due to their increased thermal inertia), locomotor activity can raise the T_b of a large, active mammal remarkably quickly—a phenomenon called *heat storage*. This term might be somewhat confusing, and to be clear, these animals are not storing heat for use later; rather they store it in the same way that we store emails from exiled Nigerian royalty in our spam folders, and for similar reasons (you don't want it and it does you no good, but you often can't rid yourself of it quickly enough). For instance, a cheetah might reach a top speed of 110 kph (68.35 mph, or over 30 m/s) over a period of usually no more than fifteen seconds (although it probably won't—see Chapter 7), and during that brief time it can burn energy up to fifty-four times faster than it does when at rest.

A 1973 study of heat balance in sprinting cheetahs by Dick Taylor and Victoria Rowntree at Harvard University showed that this increase in the rate of energetic expenditure within such a short time window produces a large amount of muscular heat, enough to raise the T_b of these animals by 1°C–1.5°C (1.8°F–2.7°F). Indeed, Taylor and Rowntree went on to show that the experimental animals refused to run at T_bs exceeding 41°C (106°F),[8] giving rise to the idea that the speed that they can achieve during hunts is limited by this rise in T_b rather than any of the other factors that potentially constrain maximum sprint speed (although a study on free-ranging cheetahs casts doubt on this notion by showing that these animals seldom approach such extreme T_bs during hunts in nature). Thomson's gazelles, by contrast, don't run as fast as cheetahs, clocking in at a comparatively leisurely top speed of 90 kph (25 m/s), but they have the remarkable capacity to store enough heat to increase their T_bs by up to an incredible 4.5°C (8.1°F) during short (eleven-minute) runs at submaximal speeds! Such high T_bs would be the death of most mammals, yet gazelles can tolerate these high core T_bs (above 42°C, or 108°F!) thanks to specialized mechanisms in the carotid artery that limit brain temperature increase to around 1.2°C (2.2°F). This impressive heat tolerance of gazelles allows them to run long distances at relatively high speeds without succumbing to heat exhaustion that would at the very least curtail the endurance capacities of other mammals.

8. Ray Huey recalls that Taylor arrived at Harvard's Museum of Comparative Zoology one day with a large cut running down the middle of his forehead, inflicted by a cheetah who took exception to an attempt to measure its rectal temperature following a run. Taylor was apparently very proud of the injury!

SIX

Shape and Form

Throughout this book I have alluded repeatedly to how the shape and size of both animals and parts of animals affect performance, all the while avoiding a discussion of the relation between form and function. This is a deliberate organizational gambit on my part, in that I want readers to become acquainted with the "why" of performance before confronting the perceived twin horrors of physiology and mechanics that illuminate the "how." But if you've survived the chapter on thermal physiology, then I hope you will be prepared to follow me deeper into the performance rabbit hole, as we consider how the various shapes of animals influence their performance abilities. Along the way, I hope to show how functional morphology opens the door to a deeper and richer understanding of the natural world, and also illustrates the ingenuity of evolution by natural selection to shape adaptations that allow animals to thrive and reproduce in the face of all manner of environmental challenges. Some of these adaptations are also really neat.

Performance abilities are the emergent outcomes of interactions among the skeletal, muscular, nervous, circulatory, and respiratory systems, all of which contribute to an individual's repertoire of athletic capabilities. Those capabilities in turn influence the individual's ability to survive and reproduce. This is a long-winded and jumbled rephrasing of one of the most influential heuristic biological frameworks of the past several decades, the *ecomorphological paradigm*. Formulated by Steve Arnold in 1983, the ecomorphological paradigm says that an individual's morphology determines its performance, and its performance, in turn, determines its fitness. "Morphology" here refers not only to an animal's skeletal characteristics, such as the shape and size of its bones, but also to the form and arrangement of those other features I list above that could conceivably fall under the umbrella of physiology as well.

Arnold's paradigm has fixed the course of performance research ever since, designating performance as the interface between an animal and its environment (and thus the target of selection) and codifying the notion that the shape, size, and form of animals evolve to meet their performance needs, not the other way around. The importance of the ecomorphological paradigm to performance research cannot be overstated, and it remains relevant more than thirty years on, even as researchers have increasingly recognized the significance of additional factors that can affect all three components.

Matters of Size

The success of the ecomorphological paradigm lies in its predictive ability. Not only can we use it to explain how animals accomplish spectacular athletic feats by looking at how those animals are shaped and built, but it also enables us to estimate what animals should be capable of based on that morphology. The influence of body size on performance is especially important because of how both morphology and function change according to the size of the organism in question. This means that we can get some idea of an animal's athletic capabilities from its size alone.

Consider sprint speed. For a terrestrial animal, speed is stride length (the distance between two successive placements of the same foot) multiplied by stride frequency (the number of strides taken in a given time). Speed can therefore be increased by taking longer strides, by taking more strides within a given time, or by a combination of the two. Most animals increase speed by modulating both, but they usually rely on increasing stride length after they reach high speeds. Because the length of an individual's legs limits maximum stride length, one prediction we can make is that larger animals (with correspondingly longer legs) will show faster top speeds than smaller animals. If we compare the maximum sprint speeds of animals that span a range of sizes, this is exactly what we find. There is an overall positive relation between body size and speed, and even elephants, the largest and heaviest extant terrestrial animals, are capable of a surprising turn of speed thanks to their long limbs and consequently large stride lengths. (Rhinoceroses, though large, aren't particularly fast because their legs are also relatively short for their size.)

Running in elephants is unusual because these animals lack the aerial phase in their locomotion that normally distinguishes running from walking. In other words, whereas other four-footed terrestrial mammals exhibit changes in footfall patterns called gait transitions (e.g., changing from walking to trotting and then from trotting to galloping) as their running speed increases, such that all four feet are off the ground briefly while trotting or galloping, elephants have at least one foot on the ground at all times. This is presumably because getting up to 6 tonnes (13,200 lb.) of animal airborne is a tall order, as well as because catastrophe and injury await an airborne elephant that fails to stick the landing. Thus elephants don't, and most likely can't, jump either. But their long limbs still enable them to attain some impressive speeds even while (technically) walking, with Asian elephants reaching a recorded top speed of around 40 kph (~25 mph). African elephants are pretty quick, too, as I learned first-hand many years ago in Victoria Falls, Zimbabwe, when I was forced to make a hasty escape from a bull elephant that I surprised while making a blind turn on a bicycle. (To be clear, I was on the bicycle; the elephant was on foot.)

Size effects on animal performance are pervasive and not limited to sprint speed. Sometimes these effects are counterintuitive. For example, smaller animals are weaker than larger animals in absolute terms but *for their size* are much stronger than larger animals. The reason lies in an important fundamental physical relation between surface area and volume. If we consider a shape such as a cube, we can calculate the surface area of that cube by multiplying its length by its breadth, which is equivalent to squaring its length, l (because all lengths of a cube are equal; thus $l \times l = l^2$) and then multiplying it by 6 (the number of faces on a cube), giving us $6l^2$. Volume is length $\times$ breadth $\times$ height ($l \times l \times l$), which gives us l^3, again because all lengths of a cube are equal. If we consider a different shape, say a sphere of radius r, we find something very similar: surface area = $4\pi r^2$, and volume $= \frac{3}{4}\pi r^3$. Ignoring the constants (6, 4π, and $\frac{3}{4}\pi$) which do not affect the relation between area and volume, in both cases we see that surface areas are some length, either l or r, squared (L^2), whereas volumes are some length cubed (L^3), and this turns out to be true for all isometric objects (that is, objects that have the same shape but different sizes). This means that if we double any length of an object without changing its shape, we increase its surface area four times (2^2), but its volume eight times (2^3); if we increase length fourfold, surface area increases sixteen times but volume sixty-four times; and so on.

This smaller increase in surface area relative to volume as size increases has important consequences for how animals function. The relevance to strength is that the force with which a muscle contracts is proportional to its cross-sectional area, and the amount of muscle an animal has is proportional to that animal's volume. So large animals are stronger than small ones because they are bigger and have absolutely more muscle than smaller ones; yet because volume increases faster than surface area as the animal gets bigger (again, volume is L^3, surface area is L^2), this means that smaller animals have greater cross-sectional muscle area *for their size* when compared with larger animals. This fact accounts for the ability of very small animals such as ants to lift objects that are many times their own weight—for instance, an African weaver ant in Cameroon was once recorded holding a dead bird approximately twelve hundred times its own mass, a feat of strength that would be impossible for a much larger animal with relatively smaller muscle cross-sectional area (fig. 6.1).

Scaling effects can affect performance abilities in other ways that are even less intuitive. Small spiders have a unique and special way of covering long distances, called ballooning or kiting. Unfortunately, ballooning doesn't exactly involve reenacting 1956's *Around the World in Eighty Days* with an all-arachnid cast and tiny top hats (which would make it infinitely more watchable). Instead, little spiders adopt a posture called tiptoeing whereby they raise their abdomens above their heads and then release one or more long, thin strands of silk (called gossamer) from their spinnerets. These strands catch the wind and lift the animal into the air. Spiders can gain impressive altitude with this method but cannot steer and are at the

Fig. 6.1. Relatively large muscle cross-sectional area, coupled with
a remarkable adhesive capacity of the feet, allows this weaver ant
to suspend a 7 g (0.25 oz.) dead bird. Groups of these ants regularly
transport very large prey items, and the remains of lizards, snakes, birds,
and bats have been found in their nests. Photo by Alain Dejean.

mercy of air currents. Nonetheless, ballooning is extremely effective; after the epic 1883 eruption of the volcano Krakatoa, which killed every living thing on that island, the very first known animal to recolonize the island was a small spider that had ballooned in from elsewhere.

The ability of spiders to balloon is size-dependent, and spiders larger than 1 milligram are likely unable to do so. This is not only because heavier spiders are harder to lift. Small spiders balloon by taking advantage of a type of drag that is mostly unfamiliar to larger animals. *Drag* is the term used for resistance that a body experiences when moving through a fluid. It acts on such bodies in two ways, and we therefore recognize two kinds of drag: *pressure drag*, which acts on the frontal area of a body moving through a fluid; and *friction drag*, which acts on the surface of the body parallel to the fluid flow direction and is a property of fluid viscosity (that is, how easily that fluid flows). We don't often think of air as a fluid in our everyday lives, but its characteristic of flow indeed makes it one. You can thus think of pressure drag as the resistance you feel on your face when you put your head out of a moving car, whereas friction drag is what makes it difficult to pull a very long rope out of the water, even if that rope itself isn't heavy. The type of drag that is most important for a given individual depends on how large that individual is, how fast it's going, the density of the fluid it moves through, and the viscosity of the fluid it moves through. These variables are described in combination by a dimensionless number called the *Reynolds number*.[1]

At high Reynolds numbers pressure drag is most important, and that is why things like jet planes, swifts, and dolphins are streamlined, because streamlining reduces pressure drag by minimizing the frontal area of the animal or vehicle in question that comes in contact with the fluid medium. But at low Reynolds numbers, friction drag and viscosity predominate. It is friction drag that pulls small, low-Reynolds-number ballooning spiders into

---

1. The Reynolds number (*Re*) is calculated as:

$$Re = \frac{lvD}{\mu}$$

where *l* is length of the animal or propulsive organ, *v* is speed of the animal, *D* is density of the medium, and μ is the viscosity of the medium.

the air along the length of their gossamer strands, in a way that would be impossible for a larger animal.

Very small animals and very large animals therefore experience the world in fundamentally different ways. Large animals such as ourselves are seldom aware of the fluidity of air until either we or it are moving at high speeds (and thus high Reynolds numbers). But for very small insects, even moving through still air is like rowing through syrup. This has several implications for how these animals are shaped and perform, and explains why very small insects like fruit flies are not at all streamlined, why their wings are shaped differently to those of birds, and also why they cannot glide.

Flying Birds and Biting Dragons

Body size strongly influences athletic abilities. But size isn't everything, and both the shape of the animal and the shape of specific animal parts have been molded by selection for specific performance functions. This relation between form and function is clearly illustrated by wing shape in birds. Flying animals take to the skies by generating *lift*, and their ability to do so is highly dependent on the capacity of their wings to act as *airfoils*. Airfoils are structures that are shaped in such a way that air moves over the top of the foil faster than it does over the bottom, creating low pressure above the airfoil. The resulting pressure difference above and below the airfoil (as well as the Newtonian reaction to the forces acting on the airfoil at right angles to the direction of airflow over it) produces lift.[2] Rather than grappling with concepts of fluid dynamics, though, we can see lift directly by blowing over the curved upper surface of a long, narrow strip of paper held up to the mouth. This causes the blown air to move rapidly over the top of the paper, which lowers the pressure on the upper paper surface. Since the pressure on the bottom remains unchanged the paper strip airfoil experiences an upthrust, or lift.

Non-Kryptonian flying animals possess wings that are shaped to take advantage of this phenomenon, and we can make predictions regarding

2. Lift is a curious phenomenon, and for it to occur, both Newton's third law and Bernoulli's principle, which says that a decrease in pressure accompanies an increase in fluid velocity, come into play.

their flight capabilities based on their wing shape. There are two relevant descriptors of wing shape that affect flight performance: *aspect ratio* and *wing loading*. Aspect ratio is the ratio of wing length to wing width, such that individuals with long, skinny wings have high aspect ratios, whereas those with short, stubby wings have low aspect ratios (fig. 6.2). Wing loading is the ratio of the mass of the animal to the combined surface area of the wings, and so individuals that weigh a lot and have smaller wing areas have high wing loadings, whereas those that have low weight relative to wing area have low wing loadings. The specific combination of wing loading and aspect ratio determines the flight capabilities of an individual: those with low

Fig. 6.2. Birds with (a) high aspect ratio wings (northern royal albatross) and (b) low aspect ratio wings (Harris hawk).
Photos by Benchill and Tony Hisgett, respectively.

aspect ratio wings and low wing loadings are more maneuverable, whereas high aspect ratio wings and moderate to high wing loadings are best suited for gliding and soaring animals. Animals with higher wing loadings also flap less in general.

Aspect ratio and wing loading have other implications for flight performance. For example, airspeed is directly proportional to the square root of wing loading when gliding at a constant speed and angle, which means that animals with higher wing loadings glide faster than those with lower ones (and consequently also lose altitude faster, because gliding involves descent by definition). A group of birds called the procellariforms, which includes gulls, shearwaters, and albatrosses, use their high aspect ratio wings to glide for extremely long distances and to practice a form of soaring (that is, gliding while gaining altitude as opposed to losing it) called slope soaring. This involves their using horizontal winds that are deflected upward by cliffs or waves to gain altitude. The wings of the maneuverable female warblers of Chapter 4, on the other hand, have a lower aspect ratio than those of the males, who are more concerned with covering long distances to reach nesting sites quickly and cheaply and thus also have higher wing loadings.

Bird wings are not static structures, and birds can exert further control over their glide performance by altering the wings' shape. Swifts extend their wings fully when gliding or turning slowly, but they sweep their wings back for fast glides or tight turns. Bats, however, exhibit a much narrower range of wing shapes than birds, probably reflecting the narrower range of habitats that bats inhabit. No bats exhibit the combination of high wing loadings and high aspect ratios found in procellariforms, and thus none can soar or glide as long or as far as seabirds. Finally, insect wings are complex structures, and the ways that small insects in particular use their wings to achieve lift are varied enough to keep functional morphologists busy for a long time to come. Wing shape and function are more familiar in larger insects, though, and there are some clear shape differences between the wings of migratory dragonflies that demand efficiency (and can be used to glide) and those that require high maneuverability for guarding females.

Generating lift via an airfoil is all well and good, but how does one go from gliding and soaring to powered flight? Engineers realized some time ago that the simplest way to build a flying machine is to divide the processes of going upward and going forward and assigning them to wings and to engines, respectively. But in nature, the wings alone do both. They do this in part by rotating the wing forward to change its angle of descent relative to the body such that the wing generates lift that is now not only upward but has both upward and forward components. This orientation accompanies the downstroke in flapping flight, which imparts rearward momentum to the surrounding air via pressure drag, while the resulting equal but opposite momentum pushes forward on the wing area, moving the animal forward. That downstroke has to be powerful enough not only to allow the animal to move forward but to continue doing so even during the recovery stroke while the wing returns to its initial pre-downstroke position.

This combination of wing orientation and flapping accounts for the incredible aerial agility of hummingbirds, which can fly forward, fly backward, and even hover by changing the orientation of their beating wings as few other animals can to direct the resultant lift in various directions as required (it also helps that hummingbirds are small). Dragonflies can control their two pairs of wings more or less independently, which grants them enormous flexibility in flight style. The wings themselves also bend and twist during flight, likely adding to their nimbleness in the air in complex ways. Small insects such as mosquitoes, however, generate lift very differently to larger ones, and researchers led by Richard Bomphrey at the Royal Veterinary College filmed mosquitoes using eight high-speed cameras with frame rates of ten thousand frames per second to figure out exactly how. Mosquitoes flap their skinny wings more than eight hundred times per second, twisting and rotating them in such a way that generates complex vortices (that is, swirls of air) and a special kind of rotational drag which together generate lift during both the upstroke and downstroke phases of their unusually shallow wingbeats. No other insect flies like a mosquito does, to the point where Bomphrey and colleagues speculate that their unique flight mechanism might be driven in part by a need to generate the characteristic buzzing sound of mosquito wingbeats for acoustic signaling purposes.

Although functional morphology overall does a good job of predicting animal performance abilities based on shape, it is important to always bear in mind the integrative nature of performance, and particularly the influence of behavior. For example, despite being the largest lizards in the world (not to mention prodigious carnivores), the 3 m (9.8 ft.) long Komodo dragon has small bite muscles, and thus weak bite forces, for its size. Dragons therefore do not perform the suffocating killing bites that other large carnivores use to subdue their dinner, and these animals were long thought to rely on a particularly unpleasant cocktail of bacteria in their saliva that they chewed with serrated teeth into their prey, leaving the victim to wander off and eventually die of sepsis. This idea was first outlined in a book on Komodo dragon behavior and ecology by the herpetologist Walter Auffenberg, who, along with his wife and children, spent almost a year on the island of Komodo observing the dragons in action, on what one assumes was either the best or the worst family vacation ever. But while Auffenberg's book is valuable in many respects, his explanation of the dragon's lethal bite is almost certainly wrong.

Komodos don't rely on super bacteria to poison their prey. Instead, they produce venom. More important, as a study by Dominic D'Amore of Daemen College shows, their massive neck and head can exert and withstand extremely large pulling forces. These facilitate the dragons' actual feeding method, which involves embedding their slashing teeth deep into prey, releasing the venom (which also contains a potent anticoagulant), and then ripping out chunks of flesh, causing the prey animal to die from a combination of shock, venom, and blood loss. In this case, the specifics of the behavior (the rotation and pulling of the head) are just as important as the animal's functional ability to enact a successful prey capture, showing that it really is in how you use it.[3]

3. The Sylvester Stallone classic *Over the Top* narrowly misses a perfect opportunity to illustrate the interplay of morphology, behavior, and performance. In this objectively amazing movie, Stallone plays one of the thousands of truck drivers/arm wrestlers who notoriously roamed the United States during the 1980s. In an attempt to bond with his estranged and awful teenage son Mike, Stallone introduces him to arm wrestling because the only way Stallone knows how to relate to other humans is through violence. Mike first loses to a larger, more hatable teenager, prompting Stallone to deliver to Mike a highly insincere pep talk about how much he believes in him. Confusingly, Stallone never bothers to explain to Mike the mechanics involved in

Fast and Furious

If the shape and size of animals determines their performance abilities, then for any given performance ability there should be a particular size-shape combination that is suited to optimal performance. Compared across species, animals that rely on different performance abilities in their day-to-day lives are indeed shaped in ways that enable those abilities, just as the eco-morphological paradigm predicts. Biologists often prefer to work on those animals, called *model organisms*, that perform some particular function exceptionally well because the underlying mechanical principles involved become more obvious and thus easier to study in animals where that function is greatly exaggerated. So by figuring out the physiology and mechanics involved in something that is, say, an exceptionally good jumper, we can apply those insights to other species that may not jump quite as well but nonetheless jump in similar ways and therefore rely on the same fundamental properties to do so. Of course, that isn't the only reason these animals are of interest. Some of them are just amazing.

If we consider pure speed, the very fastest animals in nature share some obvious features in common, such as a streamlined body shape to reduce pressure drag. But beyond the gross similarities, natural selection has tweaked the design of these animals to push the envelope of their speed limits. Thus, the very fastest animal athletes have evolved suites of speed-enhancing traits that work synergistically to produce extreme athletic performance.

The obvious animal to start with is the cheetah. Cheetahs' amazing speed can be traced directly to a number of morphological adaptations for sprinting ability. They have a muscular back and spine that flexes and then springs back into shape like a bow, driving the legs farther apart than would otherwise be the case. Not only are their limbs relatively longer than those of other cats, but the forelimb shoulder joint is also capable of more movement compared with those of similar-sized animals such as greyhounds. These fea-

---

arm wrestling, which would not only have been infinitely more useful but is also the premise of the movie. Mike inexplicably wins the rematch anyway, disappointing those of us in the audience who yearned for his failure and humiliation.

tures enable cheetahs to achieve a stride length of 6–8 m (21–26 ft.), which, at about one and a half times the height of an average giraffe, is exceptionally long. The flexible spine also assists the hindlimbs in supporting about 70 percent of cheetahs' body weight, particularly when moving at speed, and overall the spinal column and associated muscles boost their speed by around 10 percent. The muscles powering cheetah limbs are also made up mostly of a type of anaerobic muscle fiber called fast-twitch fibers that are adapted for burst power output under low-oxygen conditions, as befits a sprinter, and contribute to their blazing acceleration (roughly twice that of a horse). Their long, flattened tail acts as both a counterweight and a rudder when changing direction, contributing to their exceptional agility and maneuverability, which are arguably as important for successful prey capture as their speed, if not more so. This agility is enhanced by elongated footpads and nonretractable claws that function as running spikes, increasing traction.

Cheetahs are a marvel of evolution. But because cheetahs are so charismatic and fascinating, there is also a great deal of misinformation about them, and one doesn't have to look far to find any number of other features that are often touted as adaptations to speed. These claims mostly don't stand up to scrutiny: few are grounded in any evidence from the peer-reviewed scientific literature, and some are either misinterpretations or untested inferences of function from morphology.

One such claim is that cheetahs purportedly have relatively large hearts, lungs, and nasal cavities that supply the animal with increased oxygen during runs. However, cheetah hearts are not especially large for their size, and in any case an increased oxygen supply offers little benefit during sprint performance that does not use oxygen (although larger hearts could arguably allow swifter recovery following anaerobic exertion). Cheetahs do indeed have markedly enlarged nasal cavities, but these are far more likely to aid the animal in breathing while it uses its relatively weak jaws to slowly suffocate its prey, which it cannot do as quickly as some other big cats. Also of note is that those cavities support an unusually high density of specialized bony plates called turbinates that could possibly be beneficial for cooling the brain as body temperature rises during sprints (as discussed in

Chapter 5). But they also might not, and large nasal cavities and dense turbinates could conceivably exist for other reasons or even be a selectively
neutral consequence of some aspect of skull design.

Fast as they are, cheetahs are not the fastest animals on the planet. The
very fastest animal is the peregrine falcon. In nature these animals capture
their prey on the wing, the falcon's strike being the culmination of high-
speed dives that might begin from several hundred meters above the prey
item. Their top speed is disputed, but a diving peregrine falcon has been
recorded moving at 389 kph (242 mph), roughly equivalent to the wind
speed of an F4 tornado. To get these measurements, a team of mathemati-
cians and engineers strapped a small skydiving computer onto a tame 1 kg
(2.2 lb.) falcon named Frightful, released her from a plane to allow her to
achieve level flight, and then skydived with her from an altitude of over
5,100 m (17,000 ft.). Frightful's top speed has been challenged by some biologists on the grounds that peregrine falcons in nature are not known to
dive from such a height. However, measures of these animals' dive speeds in
the scientific literature, though slower and ranging from 185 kph to 325 kph
(115–204 mph), still place them at the top of the performance list.

Falcons such as Frightful achieve these astonishing dive speeds not only
through their streamlined body shape but probably also by controlling the
attitude of individual feathers and the contours of their bodies while in motion to achieve precise corrections in velocity. Although such speeds are
seen only during dives, some studies suggest that these birds nonetheless
normally have to fly at higher speeds than other raptors simply to avoid
stalling, similar in concept to how Formula 1 racing cars can only be driven
properly at speeds high enough to warm the tires and enable them to grip
the track surface. This is partly due to the falcon's high aspect ratio and wing
loading relative to other raptors. Indeed, peregrine falcons are heavier than
close relatives such as Lanner falcons, which aids them during dives, and
both their wing loadings and wing aspect ratios are also higher than those of
Lanners. Falcons fold their wings during dives, but as I mentioned above,
higher aspect ratios describe wings that are best used for soaring and gliding.
Peregrine falcons make use of both, soaring on thermals to gain altitude for
those high dives.

The list of fastest animals is dominated by birds, for reasons that probably have to do with the inherent differences in cost between flight and other modes of locomotion (about which more in Chapter 8) and the influence of gravity. But whereas all flying animals use variants of pretty much the same scheme, there are several ways to get around in water. One method, called jetting, is popular (albeit expensive) and relatively simple, based on the notion of conservation of momentum.[4] It also requires little morphological modification beyond a muscular tube with water input and an exit. Squid get around this way, as do jellyfish.

Swimming shares many similarities with flight, including the use of airfoils for generating lift and thrust (to the point where underwater swimming in penguins and skates has been called aquaflying), but it differs from flying due to the vastly increased density and viscosity of water relative to air. This density, however, also allows for several other ways of generating lift. Eels do so by using the undulating motion of their entire bodies to produce lift (actually thrust in this case) laterally on either side, the combined action of which propels the animal forward. Trout do the same thing, but to a lesser extent, moving mostly the tail and the trunk connected to it, whereas tuna use only their tail to produce thrust, with the trunk not being involved at all. Swimming and flying thus both boil down to imparting rearward momentum to the surrounding medium while relying on the equal and opposite momentum to overcome drag and propel the animal forward.

The fastest swimming animals are probably the billfish — a group of large, predatory fish including sailfish and swordfish. Growing to a maximum size of 3 m (just under 10 ft.) and 90 kg (200 lb.), sailfish hunt smaller and slower, yet more maneuverable, prey, deploying their characteristic sail-like dorsal fin as a brake, which increases drag and allows them to change direction

4. Momentum is mass × velocity. Squirt a jet of water out of a tube or hose, and that water leaves the tube quickly. But because water is incompressible, the same mass of water always leaves per unit time regardless of flow rate. Increasing flow rate therefore increases momentum in the direction of flow (mass is constant, but velocity increases), and because momentum must be conserved (thanks, physics!), it also increases in the opposite direction. Water therefore flows faster through narrow tubes, and this is why hoses with narrow nozzles push back more strongly than those with wider ones as the water exits.

quickly. This animal's morphology appears optimized for fast swimming, with a streamlined body shape and a powerful, muscular tail ending in a high aspect ratio caudal fin roughly the same shape as the crescentlike extended wings of a swallow, and for similar reasons: both can be oscillated at high speeds and experience minimal drag. Intriguingly, sailfish exhibit a number of bony, V-shaped scales that protrude from the skin's surface. The purpose of these protrusions is not known with certainty, but one idea is that they disrupt the smooth flow of water over the skin's surface, decreasing friction drag that would otherwise slow the animal.

Swordfish may be almost as fast as sailfish but have much smaller skin projections called denticles, so small in fact that the skin nonetheless feels smooth. Despite their size, these structures likely do aid swimming efficiency, and artificial material based on the surface structure of shark skin (which also bears denticles) increases swimming speed by 6.6 percent and reduces energetic expenditure during swimming by 5.9 percent. Swordfish skin also harbors pores connected via a network of capillaries to a specialized gland at the base of the sword. That gland secretes an oil that spreads across the animal's head and, in combination with the denticles, appears to reduce friction drag by forming a water-repelling layer of lubricant on the surface of the skin and preventing boundary layer separation. These oils are not limited to swordfish, and similar lubricating substances are responsible for the slimy texture of most fish, where they serve a similar purpose, reducing friction drag by more than 65 percent in fast-swimming animals such as the Pacific barracuda (although they probably also have other functions, such as protecting the animal from parasites).

But while the morphology of sailfish and swordfish certainly pegs them as speedy animals, the maximum speeds that each can attain is disputed. Some older studies have clocked sailfish moving at the remarkable top speed of 110 kph (68 mph), with swordfish not far behind at 97 kph (60 mph). Unfortunately, they may be too good to be true. Measuring the speeds of large predatory fish is a challenge for several reasons, ranging from the relative scarcity of the animals to the difficulty of studying them under controlled conditions. Their performance is therefore often estimated in nature, which introduces new problems. For example, a fish swimming slowly in a fast current can appear to be swimming much faster than it really is. Although scientists routinely attempt to account for such confounding variables, some billfish performance measures have filtered into the scientific literature from unscientific sources such as fishermen's estimates of how fast hooked fish unspool fishing line, which are far less rigorous.

In light of these problems, a study led by Morten Svendsen of the University of Copenhagen estimated the theoretical top speeds of sailfish based on how quickly the trunk muscles that power swimming contract when artificially stimulated. Those estimates are much lower, with sailfish potentially achieving a maximum speed of only 40 kph (25 mph). Additional independent evidence suggests that the aquatic environment itself does not allow for swimming speeds much faster than this (about which more in Chapter 7), which lends credibility to this lower estimate. Thus, although the streamlined morphology of these animals almost certainly does aid them in attaining their true top speeds, it probably doesn't allow them to move as breathtakingly fast as some believe.

Finally, although cheetahs, peregrine falcons, sailfish, and mantis shrimp are incredible and fascinating, I have to admit that the fastest movement in the natural world is not performed by an animal at all. Several species of fungi eject their spores at up to 25 m/s (90 kph, or 60 mph), which is very quick, but with an almost unbelievable maximum acceleration of up to 180,000 times the acceleration due to gravity! This means that these spores can travel more than one million times their length in one second. For comparison, the human equivalent would be a person traveling at five thousand times the speed of sound. This makes fungi the only part of the tree of life that can (1) move ridiculously fast, (2) be delicious, and (3) get you super high.

Running Far, Flying High, and Diving Deep

Because sprinting and endurance are powered by aerobic and anaerobic sources, respectively (as discussed in Chapter 5), animals that are specialized for speed differ from endurance specialists in several key ways. Human sprinters, for example, are stocky and muscular, whereas marathon runners are thinner and taller. This difference in body shape is driven by the different mechanical and physiological requirements for the two events.

Sprinting occurs briefly and over short distances. As such, it relies on large amounts of force exerted on the ground over a brief period of time—in other words, high power. This means that sprinters have to be heavily muscled, but not just in the legs as one might expect. Human sprinters run on only two legs, but they begin their races from essentially a four-limbed posture, crouching down and pushing up and forward out of specially arranged starting blocks. The high forces generated by the legs during a crouched sprint start are applied to the blocks but are also applied back from the blocks to the legs (via Newton's third law) and transmitted throughout the body as it straightens and changes orientation from the horizontal to the vertical, generating torsion in the torso that would cause it to twist as the legs are straightened. Resisting this inefficient twisting motion and keeping the resultant force vectors optimally aligned in the forward direction requires extremely strong core and upper body strength, and this is why sprinters look as if they spend as much time in the gym as they do on the track.

The mechanics and energetics required to optimize long-distance running, on the other hand, are different. Long-distance running occurs over much longer periods of time than sprinting. Rather than doing a large amount of work in a short period, endurance runners therefore do less work per unit time, yet for far longer periods. That requires less forceful strides, but a constant supply of energy, and because endurance runners always run below their maximum capacity to use oxygen (rather than above it as anaerobically powered sprinters do), that means a constant supply of oxygen.

Animals with superior endurance abilities exhibit features that aid superior oxygen supply throughout the body, such as proportionally larger hearts, lung volumes, number of red blood cells, and spleens that harbor greater red blood cell reserves. Within humans again, the hearts of trained marathon runners pump more blood per heartbeat than those of untrained individuals through enlargement of the left ventricle, which accounts for their lower heart rates. Training also increases the percentage of oxygen-carrying red blood cells in the blood (called the *hematocrit*), increasing their capacity to supply oxygen to their muscles. Aerobic training further affects muscle characteristics, with trained runners having more oxygen-powered muscle fibers as well as increased density and activity of mitochondria in muscle cells, as one might expect if those muscles are being pushed to consume more oxygen.

Importantly, though, such training does not increase muscle mass. A heavily muscled physique is detrimental to endurance runners, because in addition to their gaining little long-term benefit from the extra forces generated, they also have to spend more energy both fueling and carrying all of that extra muscle over long distances. Muscular legs are particularly expensive because of how the distribution of mass affects the human bipedal running gait. Experiments with adding extra weights to runners show that the distribution of that mass is more important than the mass itself—adding mass around a runner's midsection increased the energetic cost of locomotion over a given distance by around 8 percent, but adding the equivalent mass to the ankles increased locomotion costs by around 20 percent over the same distance. This is because the legs act like muscle-powered pendulums during locomotion, swinging backward and forward over the course of a step, and adding more weight to the end of that pendulum means expending more force to move it, which in turn costs more energy.

Animals that perform at high altitudes, where there is both less air and less oxygen, also experience oxygen supply challenges. But they also face several other trials, such as cold temperatures and, for flying animals, the need to increase power output to sustain flight in thin air. To deal with the oxygen supply issue, birds have highly efficient lungs that are structured such that air flows through them in only one direction at all times, as opposed to in and out like ours. This means that birds transfer oxygen through their lungs to their bloodstream while both inhaling and exhaling, not just during inhalation as we do. Birds are also able to deliver that oxygen to tissues such as muscle that require it more rapidly than other animals can through elaboration of the capillary system that brings oxygenated blood in close contact with those tissues.

High-flying birds can reach astonishing altitudes. On November 29, 1973, a commercial airliner sampled (that is, collided with) a Ruppell's vulture flying at an altitude of 11,280 m (36,975 ft.)—nearly 2,500 m (8,200 ft.) higher than Mount Everest! Less impressively (but only by comparison), bar-headed geese fly over the Himalayas at elevations of almost 2,700 m (9,000 ft.). To breathe the thin air at these altitudes, geese take much longer and deeper breaths, loading their greatly enlarged lungs with more of that air than other birds are capable of doing. They also have specialized red blood cells that bind oxygen much more readily than that of other birds, again aiding oxygen delivery to muscles and organs.

It may appear intuitive that deep-diving mammals would have very similar adaptations to increase oxygen supply while underwater, but in fact the challenges here are somewhat different. Mammals running very long distances or birds flying into airliners can still take air into their lungs while doing so because that air is available to them at all times (albeit at lower percentages of oxygen at high altitudes). But diving mammals are forced to make do with the amount of oxygen that they have in them when the dive commences for the duration of that dive. This means that diving mammals adjust their cardiovascular systems to meter out their limited oxygen stores to those tissues that need them the most or that can withstand low-oxygen conditions the least, such as the brain.

Consider the Weddell seal—not the deepest diving mammal in the world, able to achieve maximum depths of a paltry 2,000 ft. (slightly less than the height of the One World Trade Center building in New York City) as opposed to the 9,816 ft. (almost 3,000 m) achieved by Cuvier's beaked whale, the diving physiology of this animal is nonetheless well understood. To withstand dive durations of forty-five to eighty minutes, the hematocrit of Weddell seals is around 60 percent to 70 percent. This is about as high as it can get before the blood becomes too syrupy to flow through the blood vessels. Weddell seals also contain much larger blood volumes within their bodies compared to other mammals of similar size, which enables them to store more oxygen while submerged. But the lungs of these seals are not especially large. Indeed, they are remarkably small, with lung volumes roughly half of ours despite these animals' greater body size. What is more, the seals empty and collapse their lungs entirely while diving. They do this to avoid taking any air at all down with them, which seems odd because one might think that they would want to hold their breath during dives, as we do when diving without air tanks.

Seals empty their lungs before diving because of what happens to gasses under very high pressure, such as those occurring at large depths. At even 300 m underwater (about 1,000 ft., slightly less than the height of the Eiffel Tower), the pressure is roughly thirty times that of atmospheric pressure at sea level, and it only gets higher the deeper you go. Under such enormous pressures, the gasses that make up air become more soluble in liquid and thus dissolve directly into the blood plasma as opposed to binding to red blood cells. If that pressure were then suddenly to drop, the gasses would leave the liquid solution rapidly. Think of how champagne fizzes when you pop the cork and the pressure inside the bottle drops suddenly, causing the dissolved carbon dioxide to bubble out of solution; now think of something similar happening in an animal's bloodstream as it ascends rapidly from great depths and the crushing pressure is quickly relieved, causing the dissolved nitrogen in particular to form bubbles in the blood. Those bubbles block capillaries, disrupt blood flow, and are generally problematic. This phenomenon is better known to us as the bends, and it is the reason diving tanks do not contain nitrogen in particular in their air mixtures and also why deep-sea divers have to stop and wait at several points during their ascent to the surface. By not taking any air at all down with them, Weddell seals gain an advantage in terms of both decreased buoyancy, which helps them to reach such substantial depths, and in not dying painfully from decompression sickness.

Adapting to Performance Adaptations

Specialization for certain performance abilities is not limited to primary traits and organs, such as wing shape or heart size. Screaming through the air at high speeds or plunging deep into the ocean can have consequences for other basic biological functions as well, and animals have sometimes altered their behaviors, and even evolved ancillary adaptations, to cope with the side effects of their phenomenal athletic abilities. Tiger beetles run so fast that their visual system cannot cope with light detection at high speeds.

This is not to say that tiger beetles somehow run faster than the speed of light (although they are pretty fast, with a top speed of 2.23 kph [1.38 mph] . . . not too shabby for a 15–22 mm [0.59–0.87 in.] long insect!) but to note that their nervous systems have trouble distinguishing image motion that is a result of their own movement from motion in images that are unrelated to their own movement. In other words, if a prey item is moving at the same time that the tiger beetle is moving, the beetle can have a hard time figuring out whether the prey item is really moving (and if so in what direction and how fast) or whether it only *appears* to be moving because the beetle itself is in motion. The problem is exacerbated when these animals approach prey at an angle. Tiger beetles overcome this constraint by running intermittently, sprinting short distances and then stopping briefly to allow their vision to catch up. They also hold their antennae out in front of them to detect obstacles they cannot see.

Peregrine falcons can see extremely well, but they have other problems. Breathing at ludicrously high speeds is tough but necessary, and so these birds have specialized nasal cavities that funnel air into their nostrils, similar to the inverted cones in the intakes of some jet engines. Woodpeckers, to prevent brain injury brought on by repeatedly driving their faces into tree trunks at speeds of up to 25 kph (15.5 mph) and enduring decelerations of as much as one thousand times the force of gravity each time, have several shock-absorbing cranial adaptations. These range from spongelike bones in the skull that dissipate the pecking force to a specialized bone called the hyoid that loops around and over the skull from under the beak to just before the forehead and acts like a safety belt during impact. But strange as they are, all of these adaptations seem pedestrian compared to some of the truly unusual animal athletic abilities on display in nature.

Weird Animal Athletes

The east coast of southern Africa is stunningly beautiful. With spectacular shorelines, dazzling beaches, and comfortable water temperatures in the Indian Ocean, thanks to the warm Agulhas current that flows southwest down the coast, it is a surfer's dream and a nature lover's delight. The waters just offshore also play host to one of the most incredible spectacles in the natural world.

Between May and July, during the austral winter, a cold current originating in the vast and cold Southern Ocean flows northeast, forcing itself between the coastline and the Agulhas current. It carries within it millions upon millions of cool-adapted sardines. Unable to survive at temperatures much greater than 20°C (68°F) and trapped within the strong water flow, the sardines are forced to follow the cool current up the South African coastline toward KwaZulu-Natal, creating a slick up to 25 km long and 15 m deep (15.5 mi. × 40 ft.) and comprising shoals of more than five hundred million animals.

This is the sardine run, and it attracts predators in enormous numbers. Thousands of dolphins harry the hordes of sardines from below, blowing bubbles to corral the fish into smaller groups that they then force up toward the surface where the sardines' movements are more constrained. Sharks follow in their hundreds, taking advantage of the dolphins' hard work, as do Cape fur seals, up to a point. But once these "bait balls" of panicked fish begin to approach the surface, they also become vulnerable to attack from another kind of predator: birds.

Flying in from their nesting grounds on Bird Island, Cape gannets track the sardine run from the skies. Gannets are capable fliers, but what is really special about them is their diving ability. When they spot a bait ball, the gannets dive down to grab their share of the prey. From heights of 30 m (slightly less than 100 ft.) gannets fold their wings and, like arrows raining down on medieval battlefields, plummet head first toward the water, hitting the surface at speeds of up to 86.4 kph (53.69 mph) and plunging into the thick of the mass of fish to grab an underwater meal. Diving at such speeds requires skill; broken necks in young birds are not unheard of.

Once underwater, gannets rely primarily on their momentum (again, mass × velocity) to drive their descent. They can attain depths of up to 30 m (98 ft.), but tend to use their wings and feet to brake at shallower depths, in a bizarre underwater ballet. The birds' predatory underwater sojourns follow a U-shaped profile, using buoyancy, bolstered by air trapped beneath their feathers, to return to the surface. Less frequently, underwater gannets flap their wings as if in flight to descend again, aquaflying like penguins, probably to target another fish after a failed first strike. The bait balls are short-lived, but while they last the surface of the ocean appears to boil as dolphins and sharks breach the surface in pursuit of the doomed sardines while gannets dive from the skies to the water in their hundreds, leaving trails of bubbles behind them like fireworks exploding in reverse.

The sardine run used to be an annual occurrence but has become sporadic, likely due to the warming of the earth's oceans and the consequent decrease in predictability of the cold current. However, animals that frequent habitats or media unusual for others of their kind exist elsewhere as well. Several families of flying fish do the opposite of gannets, swimming to the surface and leaping from the water to glide on their elongated, airfoil-like pectoral fins.[5] Some species even use four fins for flight rather than two, having evolved elongated pelvic fins as well. The submerged four-wingers approach the surface at high speeds and shallow angles (36 kph/22.4 mph, and about 30 degrees to the horizontal) and are catapulted forward almost entirely out of the water as they break the surface and enter the eight-hundred-times-less-dense air. However, the rear part of the stiffened tail remains in the water and beats vigorously at fifty to seventy strokes per second for a thirty-second or so taxi flight before it clears the water entirely into free flight. These fish can glide at airspeeds of 54–72 kph (33.5–44.7 mph) and a maximum altitude of 8 m (26 ft.) for distances of around 50 m (164 ft.) before airspeed falls and the tail is lowered back into the water for another taxi flight. In this way, four-wingers can achieve total flight distances of up to 400 m (1,300 ft.) before reentering the water.

There is a lot of variation in flight among species of flying fish, much of it related to size. Some forgo the taxi flight entirely, especially when taking off out of fast-moving waves, and others soar off of waves as procellariforms do. The fins of these fish are well adapted to do so, and some of the larger species exhibit wing loadings and aspect ratios comparable to those of glid-

5. Despite their common name, these fish only glide. Nonetheless, a controversy dating back at least to the mid-nineteenth century centered on whether these animals exhibited powered flapping flight. This controversy was laid to rest finally in 1941 by film of these fish in action that showed unequivocally that they do not.

ing birds. Somewhat surprisingly, why this exceptional ability has evolved in these animals is poorly understood, although potential explanations include predator escape and energetic efficiency.

Fish may be the strangest animals to have evolved gliding abilities, but they have some stiff competition in this regard. Gliding has evolved independently more than thirty times in a number of traditionally nonflying animals, from mammals and fish to reptiles, amphibians, and even a species of squid that uses water-jet propulsion to launch itself out of the water, at which point it unfurls its airfoil-like fins, allowing it to glide up to 30 m in about three seconds!

The better known of the sixty species of mammalian gliders such as sugar gliders and colugos have large flaps of skin called patagial membranes that extend laterally between the limbs, and sometimes also between the limbs and structures such as tails, to form the gliding surface. Others, such as the marsupial feathertail glider, rely on flattened tails that serve similar aerodynamic functions. Tree-dwelling lizards of the genus *Draco* in southeastern Asia exhibit drastically elongated ribs that they use to support patagial membranes of their own (fig. 6.3). The *Draco* patagia are often colorful and can be deployed or folded against the sides of the body like a fan, but when extended they allow these animals to glide from tree to tree at speeds of up to 27.4 kph (17 mph).

Although the remarkable rib patagia are found only in *Draco*, many reptiles exhibit gliding abilities and morphological features that enable these abilities, all of which work by increasing the aerodynamic surface area of the animal. These range from the gliding geckos in the genus *Ptychozoon* that glide with the aid of flaps of skin on the limbs, feet, head and abdomen, to *Holaspis* lizards that have fringed toes and tails serving a similar purpose. Several frogs have been accused of gliding, and many frog species have exaggerated webbing between the toes that could well facilitate glides. Some even have skin flaps like the gliding geckos and clearly are good gliders.[6] But

6. The last known Rabb's fringe-limbed frog, *Ecnomiohyla rabborum*, died in captivity in Atlanta in September 2016. This canopy-dweller reportedly used its fringed limbs to glide among trees, which means there is now quantifiably less wonder in the natural world.

Fig. 6.3. *Draco spilonotus* with extended patagia. Note also
the extended anolelike dewlap. Photo by A. S. Kono.

others, such as the Caribbean species *Eleutherodactylus coqui*, could just
as well use that webbing not for true gliding but for parachuting, whereby
descent is slowed but not necessarily controlled.

Perhaps the most unusual glider of all is the Paradise flying snake, *Chrysopelea paradisi*, which, undaunted by its utter lack of limbs and patagia,
nonetheless launches itself from treetops with abandon by first hanging beneath a branch in a J-shape and then accelerating the body forward and
upward. By straightening and flattening its body at the beginning of these
jumps, the snake roughly doubles its body width, thereby turning its entire
body into an aerodynamic surface. Following this (while still falling through
the air), the snake adopts an S-shaped posture and undulates from side to
side as if it were moving on land. By orienting the front part of its body, this
snake can turn in the direction of its head movement without banking as
other gliders do.

The undulating postural movements in the air are different both from those seen in conventional terrestrial snake locomotion and from those of swimming snakes. They are also distinct from more unusual snake movements such as sidewinding, which works by fixing part of the body on the sandy ground and then lifting, moving, and setting down an adjacent part. The part of the body right next to that which was just set down is then itself raised and moved. This propagates waves of bending down the length of the animal's body and causes it to move, leaving indentations in the sand at right angles to the animal's direction of motion. By contrast, it is not entirely clear what purpose the aerial undulations of the Paradise flying snake serves, if any. This curious mode of locomotion is nonetheless effective: these snakes achieve glide performance comparable to, and sometimes exceeding, that of other, more obviously specialized gliders.

Evidence from other animals also shows that patagia are not always required for gliding, although they certainly make it easier, and some animals affect their descent via postural alterations alone. Even several species of ants control their aerial trajectory when falling from trees! An especially interesting finding is that some arboreal frogs (and even nominally terrestrial geckos) that do not exhibit any specialized gliding morphology nonetheless automatically adopt skydiving postures under conditions of microgravity that increase their surface area, thereby potentially increasing their gliding and parachuting capabilities. Microgravity also affects insect aerial performance in interesting ways: moths don't fly but prefer to float; houseflies fly not very well and prefer to walk on walls; honeybees can't fly at all and sort of tumble around. If you are wondering how we know all of this, it is because we have sent some of those animals up into space and seen them do it, while others were observed in artificially induced microgravity during parabolic flight in an airplane, just as astronauts experience when training aboard the Vomit Comet.

You might well imagine that gliding is the weirdest thing that fish are known to do. Consider, however, the small goby that occurs in the waters off Hawaii. Just as salmon feel the irresistible urge to return to their birthplace, so gobies are compelled to travel to their preferred spawning areas in the rivers of the Hawaiian Islands. Unfortunately for the gobies, the islands are mountainous, and where you have rivers and mountains, you have waterfalls.

Gobies are far smaller than salmon, and thus they are unable to pull off the spectacular leaps that salmon perform as they make their way up-river. Many waterfalls stand between the gobies and their adult spawning grounds, some as high as 350 m (again, roughly the height of the Eiffel Tower), and so the gobies do the only thing they can—they climb them. At least, they climb the rocks behind them, and they do so using pelvic fins that have been modified by natural selection to form suction cups.

Some goby species climb painfully slowly, inch by inch, alternating between the pelvic cup and a second, oral suction cup, and are thus known as inchers. By contrast, the so-called powerburst climbers are less patient and move upward by rapidly undulating their bodies from side to side, jumping vertically from attachment point to attachment point up vertiginously high cliffs. Both inchers and powerburst climbers stop for rest periods, although powerburst climbers rest for longer. They make up for this by moving faster when they aren't resting, to roughly the same net effect in terms of distance. As always, there are size effects on waterfall climbing performance, with smaller individuals tending to be better climbers than larger ones, for reasons that I have mentioned earlier.

Meanwhile in Thailand, a species of blind cavefish has been documented climbing waterfalls in a much more familiar manner. Brooke Flammang of the New Jersey Institute of Technology and colleagues found that this animal not only climbs but can move about on land as well. This in itself is not terribly shocking, as several fish have independently evolved terrestrial locomotion. For example, some eels and sticklebacks flop across patches of dry ground using the same motions that they already use for swimming, while mudskippers employ the more sophisticated method of crutching, which involves using their tails and rear part of the body to vault themselves forward and over their pectoral fins. What is noteworthy about these cavefish is that the walking gait they use is strikingly similar to that used by terrestrial vertebrate animals, especially salamanders. The pelvic morphology exhibits a number of features that have thus far only been observed in terrestrial vertebrates and are unusual in fish. These animals therefore offer a rare glimpse into how fish may have moved when they first invaded the terrestrial environment.

Several animals are also able to make a living by walking or running over water. Small insects such as water striders take advantage of a property of water called surface tension, which is the tendency of adjacent water molecules to strongly attract each other, to walk across water without exerting enough force to overcome that surface tension and break the surface. Even some tiny geckos get by using the same trick. Fishing spiders are the largest invertebrates to use surface tension to move across the surface of water, but they also adopt a running gait that involves leaping both off and back onto the water surface. Some animals take it a literal step further, though, taking several successive strides and running over the water surface.

Running on water appears impossible, because we know from our own experience that we need to push down on a firm surface to have that surface push back on us (via Newton's third law) with sufficient force to enable locomotion; this is why running on a sandy beach, which doesn't return all of the force we impart to it, is harder than running on solid ground (and why sidewinders evolved their unusual gait). Basilisk lizards nonetheless are also able to run bipedally across water using only their hindlimbs, but they are far too large to rely on surface tension to do so. They instead employ a slapping gait, using their enlarged feet to strike the surface of the water very quickly.

Slapping and rapidly withdrawing the foot before the water closes over it allows basilisks to take advantage of the water's tendency to resist displacement (familiar to anyone who has performed a belly flop) to partially support their weight until the next step. They need to pull this off only for the split-second (~0.06 s) that each foot is in contact with the water surface, and the faster the basilisk's foot displaces (or moves) the water, the more that water resists being moved (the relevant physical concepts at work here are impulse and conservation of momentum). Big feet are important, for two reasons: first, they impart pressure drag (which is a function of both the surface area of the foot and the speed at which it is moved) on the water itself; and second, while the foot is at the bottom of the (very temporary) air-filled space that it creates by pushing down on the water surface, the water on the underside of the animal's foot exerts upward force on that foot equivalent to the hydrostatic pressure of the water at that depth (hydrostatic pressure is depth × density of water × gravitational acceleration) multiplied by the surface area of the foot. To put it plainly, *the greater the surface area of the sole of the foot, the more total upward force that foot experiences.* That upward force, in combination with the water resisting entry of the foot, is enough to keep the animal afloat with each rapid step and allow it to run across its surface. After withdrawing one foot, the other is then immediately slapped down ahead of the animal's center of mass, and the process repeats. As you might imagine, there is an upper size limit on the effectiveness of this strategy, which is why water-running is usually limited to juvenile basilisks. I was fortunate enough to startle a young basilisk into running across the surface of a stream in Costa Rica several years ago, and it really is a sight to see!

Grebes, a kind of water bird, are much heavier even than basilisks, yet they exhibit a similar water-running behavior called rushing—best described as a synchronized male-female dance across the surface of the water—as part of their courtship ritual. Grebes, too, use a type of slapping locomotion, similar to the basilisk's but with several differences in gait, that generates enough force to support their larger mass. In essence grebes do what basilisks do, but harder and faster, and are thus able to accomplish a truly astonishing animal athletic feat.

The list of animals doing outlandish or unusual things is long. It is always important to bear in mind, however, that natural selection doesn't deal in whimsy, and whenever we see an animal doing something that we might consider to be out of the ordinary, there is always a good reason. It is also important to remember that even though evolution can produce some amazing things, it also has its limits.

SEVEN

Limits and Constraints

In issue 8 of the 1985 DC Comics comic book series *Crisis on Infinite Earths* (written by Marv Wolfman and beautifully drawn by George Pérez), the Flash, the Fastest Man Alive, died heroically saving the multiverse. This was Barry Allen, the second Flash, an important character whose first appearance in 1956 kicked off the lysergic Silver Age of comic books. The Flash had always been overpowered, moving fast enough to run vertically up buildings, generate tornadoes by running in circles, and even vibrating his individual molecules fast enough to cross dimensions, and one sometimes felt that his considerable abilities were wasted dealing with gimmicky villains like Captain Boomerang and the Trickster. And he wasn't just fast over short distances; he could maintain his speed seemingly indefinitely. In short, the Flash seemed able to do anything. But in this story he pushed himself further than ever, exceeding the speed of light itself to chase and outrun a tachyon (a theoretical faster-than-light particle) powering a universe-shattering weapon. In doing so, he moved so fast that he traveled backward in time and became the very lightning bolt that bestowed super speed upon his younger self. The Flash's demise lasted longer than most in comic books, and he was eventually resurrected in 2009. Although he was my favorite character as a child, the ending to his story was so metal that I wish he had stayed dead.

In real life, being struck by lightning that is also you from the future probably won't grant you super powers. Furthermore, unlike the Flash, the athletic abilities of real animals are subject to the laws of physics and the principles of mechanics and physiology. Animals cannot do everything, and as effective as selection is, various factors, ranging from the nature of the evolutionary process and the historical contingencies of how it has operated on an organism's ancestors to the mechanical properties of organismal design, constrain what is and isn't possible in terms of specific adaptations in certain animals. A further consequence of these constraints is that adaptations are often imperfect—on close inspection, living organisms are seldom as optimized for specific ecological tasks as they may appear. In this chapter, I'll mention some of the limits that apply to the evolution of athletic capacities in animals as a result of these various constraints, as well as some of the ways that those limits have been superseded.

Functional Trade-Offs, or Why Animal Performers Can't Have It All

Selection has shaped animals to perform ecological tasks that maximize fitness such as capturing prey, escaping becoming prey, or finding a mate, and most animals are required to perform many varied such tasks throughout their lifetimes. These different tasks frequently require different kinds of performance, which can place disparate and conflicting demands on the underlying individual physiology and morphology. Performance abilities are almost always compromises based on trade-offs with other elements of the musculoskeletal system that are shaped for other, different performance abilities. Consequently, an important trend throughout the animal kingdom is that excellence in a particular trait comes at the cost of less-than-excellent performance in something else; in other words, whereas it is possible to be outstanding at something, and maybe even very good two or three things, it is impossible to be exceptional at everything.

Anolis lizards, to which I have turned time and time again throughout this book, clearly illustrate not only the fit between an organism's shape and its required performance in its environment but also how that fit can be suboptimal if the organism finds itself in another environment to which it is less well adapted. Anoles are the prime exemplars of an evolutionary process called *adaptive radiation*, whereby lots of species have evolved from a single ancestral species to adapt to different ecological lifestyles in a short period of (evolutionary) time. Anole species differ from one another in a variety of ways, and those in the Caribbean show particularly interesting differences with regard to habitat preferences.

We can divide anole habitat into several categories, such as bushes, open ground, tree trunks, tree canopies, and so on. The species occurring in each type of habitat on the islands of the Greater Antilles in particular are termed *ecomorphs* because they exhibit specialized morphology and performance abilities suited to dealing with those specific environmental challenges. The six anole ecomorphs are superbly adapted to their specific ecological niches. Grass-bush anoles, for example, exhibit slender, gracile bodies and long tails suitable for "swimming" through long grass, whereas trunk specialists have splayed, crablike limbs and sprawling postures for scuttling on and around broad tree trunks—which they do with surprising alacrity. Twig anoles are not only cryptic and slow-moving, looking very much to even the careful observer like the twigs on which they reside, but their morphology is perfectly suited to moving along very narrow substrates (like twigs). Thus, twig anoles have limbs that are short for their body size because stubby limbs are more useful than lanky ones for crawling along narrow twigs and branches, probably because they allow the animal to grip and maintain its center of balance better than longer limbs do.

The morphology-to-performance relation for Caribbean anole locomotion is clear: species inhabiting broader substrates, such as tree trunks or the ground itself, have longer legs, whereas species that typically move on narrower substrates have shorter legs. So what happens when a long-legged species finds itself on a narrow substrate? If you think (or look) back to the section on rapid selection in Bahamian brown anoles in Chapter 2, you already know the answer to this question, but the experiments required to demonstrate this phenomenon were conducted in the late 1980s by Jonathan Losos and Barry Sinervo in four species of Caribbean *Anolis* lizards that varied in limb length relative to their size. After measuring sprinting ability on substrates of similar diameter to that which each species most commonly uses in nature, Losos and Sinervo then switched those species over to substrates of a different diameter. They found that as substrate diameter decreased, all species got slower, but those with the shortest limbs, although never that quick to begin with, nonetheless decreased in speed the least.

These two designs are mutually exclusive: one cannot have both long limbs and short limbs at the same time. This means that no anoles are suited for moving on both broad and narrow substrates. Long-limbed lizards that venture out onto the slender branches and twigs of their arboreal homes experience a true decrement in maximum performance ability whenever they do so, whereas short-limbed anoles cannot take full advantage of broad substrates. A subsequent experiment in a larger sample of eight anole species by Duncan Irschick and Losos not only confirmed the short-limb–long-limb substrate trade-off but also showed that long-limbed species that rely on sprinting to avoid detection or capture prey are particularly sensitive to substrate type (they tend to fall off narrow substrates a lot) and avoid those substrates that most impair their performance abilities when moving undisturbed in nature.

Such is the fit between their shape and their environment that anoles' maximum abilities can be achieved only in the habitat to which they are molded. But Caribbean anoles also illustrate a more general point regarding the evolution of integrated functional systems such as those determining performance. Evolution through natural selection is more than equal to the challenge of equipping animals for the various performance tasks that they have to undertake to ensure survival and reproduction (no matter how unusual some of those tasks may be). But sometimes there are only so many avenues available for selection to take. If two roughly similar animals in two different areas find themselves facing more-or-less similar functional challenges, natural selection may change those animals in similar ways. This is what has happened in the Caribbean anoles. Most ecomorphs occur on multiple Caribbean islands, and if we take a close look at the distribution of ecomorphs across those islands, and at the evolutionary relations of those ecomorphs, we find that any two twig anoles, for example—say, A. *valencienni* from Jamaica and A. *angusticeps* from the Bahamas—look and behave very similarly even though they are not close relatives. In fact, A. *valencienni* is much more closely related to both the brash and longer-limbed Jamaican trunk-ground ecomorph A. *lineatopus* and the large and beautifully verdant Jamaican crown-giant A. *garmani* than it is to its Bahamian doppelgänger on another island more than 650 km (400 mi.) away.

This pattern repeats throughout the Caribbean, with the same eco-morphs on different islands bearing striking similarities in morphology, behavior, and performance despite not being close relatives. Anoles are therefore also prime examples of *convergent evolution*, whereby selection has acted independently in similar ways on different species to drive the evolution of different species that nonetheless look and act very much the same. Because these species are so alike, we find that all members of a particular ecomorph category experience the same habitat-substrate trade-offs. Evolution has therefore reproduced the same fit between morphology and environment multiple times and in so doing has recapitulated the same functional trade-offs.

The *Anolis* convergence is spectacular because it has occurred in a single large genus of charismatic and fascinating animals. If we take a step back and look at evolution on a broader scale, though, we see that convergence is rampant, but we also see that it doesn't always happen in the same way as it does in anoles.

Out of the Past

The functional solutions to particular ecological challenges that evolution has come up with, while often similar, are not always the same. Specifically, the performance may be roughly equivalent, but the details of how that performance is achieved can differ, involving varying combinations of limb lengths, muscle size, and muscle attachment points. Biologists call this phenomenon *many-to-one mapping*. Put another way, different species of anoles can converge on the same performance solutions because they inherited the same functional and physiological machinery from their ancestors, but other types of animals that need to perform certain tasks may find that they have inherited from their immediate ancestors the wrong morphology for the job.

Here, natural selection must improvise. If we consider the gliding animals introduced last chapter as an example, we see that although gliding always takes advantage of the same general principles, such as airfoil-like structures for generating lift and postural adjustments that allow for control of descent, the actual gliding structures differ greatly. Mammals such as sugar gliders and colugos have their large, membranous patagia strung between their limbs, whereas *Draco*'s patagia are supported by extended ribs. Flying fish do not use patagia but have instead evolved structures that resemble high aspect ratio wings through modification of their pectoral fins. But flying fish, flying lizards, and sugar gliders are not close relatives—they are separated by millions of years of evolution and do not share a recent common ancestor from which they could all have inherited a common gliding morphology. Instead, these animals have converged on a similar performance ability. All of them can glide, but each has arrived there from different ancestral starting points through strikingly diverse functional means, pushing the limits of, but still operating within, the constraints placed on them by their evolutionary history. Yet because evolution is the ultimate pragmatist, the one characteristic that all these convergent traits have in common is that they work—sometimes spectacularly.

Convergence exposes natural selection's method of taking whatever is available and modifying it as required. Evolution is a tinkerer. It acts less like an engineer with a blueprint, unlimited funds, and an Amazon Prime account and more like a mad scientist raiding a junkyard and making it up as he or she goes. This means that the functional solutions that animals evolve to accomplish similar ecological tasks can be surprising, but it also means that sometimes the constraints are insurmountable. An animal's evolutionary history and the morphology and physiology it inherits limit its future performance prospects just as readily as it enables them, depending on which way the winds of selection are blowing.

Legs or Wings?

The problems posed by conflicting locomotor requirements are not exclusive to anole ecomorphs. Because of the clear mechanical trade-offs in optimal design, animals that attempt simultaneously to maximize multiple types of incompatible performance are almost always doomed to failure and, ultimately, extinction, which is probably why we don't see very much of them. Far more common are compromise morphologies that settle for doing one or (rarely) two things very well, while doing others at levels that may be just barely sufficient to avoid the hammer of selection.

Trade-offs and compromises are built into the overall morphological architecture of multitasking animals. Just such a basic architectural trade-off occurs between the relative size and development of legs and wings in birds. Flying birds face several aerodynamic challenges imposed by the laws of physics, and it is generally advantageous for them to be as light as possible because the less mass they have to move against the action of gravity, the better. Legs are heavy, and so birds that spend a great deal of time on the wing or that need to be highly maneuverable would do well to reduce leg size. They can't dispense with them entirely, though; having legs is pretty much mandatory for birds no matter how great a flier you are.[1] This trade-off is exactly what we see across all birds, with species that forage on the wing (and thus invest more in wings and flight) tending to have not only smaller legs but also reduced leg performance. Yet even swifts, which remain continuously aloft for ten months during their annual migratory nonbreeding period, feeding and even sleeping in flight, still have legs, albeit small ones that reflect the need for flight efficiency.

Birds are especially prone to functional conflicts because of their tendency to split their time among different locomotor modes. Penguins are splendid swimmers, but their walking abilities on land suffer as a result, being both slow and inefficient, and they are not nearly as good at tap dancing as *Happy Feet* would have you believe. Several other sea or water birds also exist in the functional limbo between two means of locomotion. Cormorants, a family of about forty species of aquatic birds, are fish eaters. Fish are scarce in the terrestrial and aerial environments frequented by most birds, though, and so cormorants have found it useful to insert themselves into the fishes' native medium instead, much as the dive-bombing Cape gannets from Chapter 6 do. But some cormorants are even better at prolonged dives

---

1. The first specimens of birds of paradise brought back from Papua New Guinea did not have legs (or wings for that matter) because those specimens were stuffed by the natives who removed the limbs before trading the rest of the bird to European explorers. This is reflected in the scientific name of the type species, *Paradisaea apoda* ("legless bird-of-paradise").

than gannets, descending as deep as 45 m (148 ft.), and have evolved features to help them do so, such as webbed feet, stubby wings, and large body size that allows them to store more oxygen relative to the rate at which they use it than could a smaller bird.

Although the Galápagos cormorant has gone all-in on diving for prey-capture purposes and has entirely lost the ability to fly, other cormorant species can still fly, albeit poorly. One reason why the diving-flight compromise morphology doesn't make for great fliers when it is heavily tilted toward the diving component is that although being heavy is great for plunging underwater (because it allows for greater momentum), in the absence of high aspect ratio wings the airborne animal has to spend more time and energy generating lift through powered, flapping flight as opposed to gliding.

Muscle Properties and Trade-Offs

Muscle is the engine that drives almost all animal movement (and thus performance). As such, its nature and design can drive performance trade-offs as well. How muscle works has implications for the types of performance it can enable, with the specific combinations of muscles, tendons that attach muscle to bone, and the bones themselves custom-designed by selection to meet the performance requirements of the animal in question.

Several facets of muscle construction, from the length of the muscle fibers to the molecular machinery that enables those fibers to contract, affect what a muscle is and isn't capable of. For example, long muscle fibers produce less force per contraction than shorter ones (all else being equal) but can also shorten in length through a greater distance than those fibers that were already short to begin with. This means that muscles comprising long muscle fibers are well suited to control and produce large, fast movements, whereas smaller but more forceful movements are controlled by musculoskeletal architecture involving short muscles and muscle fibers. Again, the fundamental incompatibility of these two arrangements is the key to understanding mechanical performance trade-offs: muscle fibers can be long and contract quickly, or they can be short and contract forcefully, but they cannot be and do both at the same time. These two properties can, however, be balanced, with the most powerful muscles (that is, those that do the most work in the shortest amount of time) being those that are intermediate for both contraction force and contraction speed.

A second property of muscle that can influence performance ability is the type of muscle fiber making up the locomotor muscles. The variety of muscle fiber types can be broadly categorized as either oxidative fibers that are rich in mitochondria, require oxygen, and contract slowly but are highly resistant to fatigue (as used by endurance running animals) and non-oxidative or glycolytic fibers (of the type favored by reptiles) that have few mitochondria, do not require oxygen, and contract quickly. Most animals have both fibers in their muscles, but animals that rely on speed tend to have a greater proportion of glycolytic fibers, and there is a direct, positive relation between sprint speed and proportion of glycolytic fibers in the locomotor muscles of mammals ranging from bears to caracals and cheetahs. The trade-off is that these faster animals have correspondingly poor aerobic capacities because a higher proportion of one fiber type necessarily precludes a large proportion of the other.

Animal performance abilities are an emergent property of muscle function. The details of how the muscles work are therefore manifest in the design and operation of the whole organism. This is apparent when we look at the muscular and skeletal architecture of different performance specialists. Greyhounds and pit bulls are both the result of many generations of intense selective dog breeding for specific performance abilities: greyhounds for running speed and pit bulls for fighting ability (although such fights are now rightly outlawed almost everywhere). The different morphological requirements for running versus fighting are rooted in the skeletal-muscular architecture of their limbs in particular. As runners, greyhounds require long limbs that they can move quickly (remember, speed = stride length × stride frequency) over distances on the order of 500 m/0.3 mi. This means that their locomotor muscles must maximize their power output. To this end, greyhounds have little muscle in their limbs, with most of their locomotor muscles being arranged at the shoulder joint. This makes their limbs lighter and easier to move at high speeds. Those muscles also have fiber lengths that both place them right in the sweet spot for power production and are also longer than those of mixed-breed dogs. Like cheetahs, grey-hounds are also rear-wheel drive animals in that they have more muscle mass in their hindlimbs as opposed to their forelimbs, which are mostly used for deceleration. By contrast, pit bulls need to both accelerate and decelerate repeatedly and explosively while maneuvering around another dangerous animal, and they require strong limbs to manipulate their opponents. This makes them more akin to four-wheel drives, with a more equal distribution of muscle across all of their much shorter limbs.

Once again, we are struck by the fundamental irreconcilability of these two animal shapes; the arrangements of both muscle and skeleton shape that make a speedy greyhound are in direct opposition to those that make a stocky pit bull. Greyhounds and pit bulls may in fact embody a general evolutionary principle: that of a trade-off between fighting and locomotion driven by the conflicting functional requirements of each.

Overcoming Constraints and Minimizing Trade-Offs

Functional trade-offs and constraints are facts of life for all animal species, but specialization for a specific athletic feat does not always preclude specialization in another, provided that the functional requirements of the two are aligned. Among the waterfall-climbing gobies that I mentioned in Chapter 6, one species of incher uses its specialized oral sucker cup not only to stick to surfaces and inch upward by alternating attachment with the pelvic sucker but also to scrape small organisms from the surface of rocks during feeding. The oral movements involved in rock climbing and rock scraping in this creature are similar, involving the same muscular and morphological architecture. Not only is there no conflict between inching and scraping, but it is very likely that one evolved from the other, although it is currently unclear which evolved first. In some species, however, selection has acted to overcome true functional constraints, or at least to minimize the mechanical and muscular trade-offs.

As we have seen, the requirements for specialized sprinters and long-distance endurance athletes are different. For the vast majority of animals, they are also opposite. If we consider human athletes specifically, one cannot be simultaneously heavily muscled for power output and slim for endurance running, nor can one have muscles made up of mostly fast-twitch, nonoxidative fibers without simultaneously decreasing the percentage of slow-twitch, oxidative fibers. Speeds that can be sustained by aerobic performance are lower than those supported anaerobically (less than half on average in mammals), which means that animals can either run very fast or for a very long time but can seldom do both.

In human track athletes, the transition from sprint to endurance running occurs somewhere at the 600–800 m range. This is why many athletes might participate in 800 m and higher races or 400 m and lower ones, but a 400 m and 800 m double is rare — indeed, no one has held the 400 m and 800 m world records simultaneously since Rudolf Harbig in 1939. Because of these different requirements, researchers interested in animal athletic trade-offs often set out with a sprint versus endurance trade-off in mind. Sometimes such a trade-off is found; for example, a study of twelve species of European lacertid lizards found that species that are excellent sprinters have poor endurance abilities but long hindlimbs for their size, and vice versa. But other studies on a variety of species (including a subsequent one on seventeen lacertid lizards) have found weak, mixed, or no evidence for such a trade-off.

The reasons why expected trade-offs might not be observed are varied and complex, and in some cases researchers require specialized statistical techniques to recover those trade-offs. However, the trade-off between speed and endurance is almost certainly real, and one species in particular offers clues as to how this particular trade-off can be overcome.

If there were an award for the best overall athlete in the animal kingdom (and if all animals had evolved the capacity to care about such things), it might well go to the North American pronghorn. Almost alone among animals, the pronghorn is able to excel at both sprinting and endurance running. With a top speed of almost 100 kph (62 mph), it is the fastest land animal that isn't a cheetah. But incredibly, the pronghorn can also maintain breakneck speeds over distances that far exceed those at which cheetahs are capable of running. There is a field observation of a pronghorn running a distance of 11 km (almost 7 mi.) in ten minutes, an average speed of 65 kph

(40 mph), and there are anecdotal reports of pronghorns outrunning pursuing light aircraft. For all this, pronghorn locomotion has been surprisingly understudied, and we know little about how these animals achieve these remarkable feats. But we do know some things.

In 1988, Stan Lindstedt (now at Northern Arizona University) and colleagues put pronghorns on a treadmill and measured their oxygen consumption while running (fig. 7.1). The rationale here is that all animals have a maximum endurance speed that is limited by their capacity to use oxygen—any faster than that speed, and locomotion must be fueled anaerobically. They set the treadmill to 36 kph (22.7 mph) at an 11 percent incline, which means that these animals were running slightly uphill at roughly the top speed of an Olympic sprinter! Lindstedt and company found that pronghorns have a VO_2max (the technical term for maximum rate of oxygen consumption) that is more than three times higher than would be

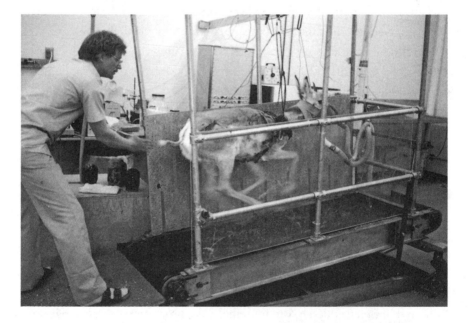

Fig. 7.1. Stan Lindstedt measuring a pronghorn on a treadmill in Laramie, Wyoming, in 1988. If you listen carefully, you can hear Senator Coburn screaming. Photo courtesy E. R. Weibel and S. L. Lindstedt.

predicted based on their size alone.[2] The reasons for this exceptional capacity are hinted at by an earlier study of the factors affecting oxygen delivery in pronghorns. Compared to an equivalently sized goat, pronghorns have a VO₂max that is five times higher; lungs twice the volume; a 50 percent greater amount of blood, with 33 percent more red blood cells and three times greater cardiac output (that is, amount of blood pumped by the heart per minute), resulting in more oxygen being distributed around the body five times faster; and two and a half times the volume of muscle mitochondria. Furthermore, pronghorns have exceptionally slim lower limbs, as befitting an animal striving to conserve energy while running long distances.

All of these adjustments point toward a species that is well adapted for endurance running, but none of them are adaptations for sprinting. So how is it that pronghorns have been able to minimize the trade-off between sprinting and endurance that we see elsewhere? And if pronghorns manage to have both speed and stamina, then why can't anyone else?

In fact, some other animals have been able to reduce the trade-off between sprinting and endurance as well, albeit to a lesser extent. The southern African black wildebeest is not a close relative of the pronghorn but has nonetheless also evolved the ability to run at high speeds for long distances, with a top speed of 70 kph/43.5 mph. This likely aids its nomadic existence across large home ranges, although this species does not participate in the spectacular mass migrations that characterize its larger cousin, the blue wildebeest. Investigations into the locomotor muscles of black wildebeest show something very unusual: high proportions of a version of fast-contracting, glycolytic fibers (called type IIx fibers) that are modified to have both high oxidative *and* nonoxidative capacity! A subsequent study showed that springbok (with a top speed of about 90 kph/60 mph) also have a lot of these modified oxidative type IIx fibers, suggesting that they may also be able to maintain high speeds over long distances. Thus, locomotion in black wildebeests and springbok appears to be enabled by muscle fibers that have overcome the trade-off between fast contraction speed and fatigue resistance

2. A recent study of aerobic capacity that included more animal species in the predictive model adjusted this value downward somewhat but still showed that pronghorns are clear outliers in terms of their VO₂max.

seen elsewhere. It is very likely that pronghorns possess these muscle fibers, too, although this remains untested.

I cannot answer the question I posed above as to why pronghorns are both champion sprinters and runners, but I can offer a surmise. The available evidence points to the aerobic capacities of pronghorns being turned up to eleven—far more so than in other animal athletes. R. McNeill Alexander pointed out in a commentary (to which he gave the excellent title "It May Be Better to Be a Wimp") of the pronghorn study that none of those adjustments to oxygen supply are evolutionary novelties but are instead exaggerations of normal mammalian structure and function. It may be that no other animal has ever been pushed to the limits of both its endurance and sprinting capacities like the pronghorns have, and that in response to strong selection pressure on aerobic capacity these animals have boosted their maximum aerobic speeds and fatigue resistance to unheard-of levels, such that they are able to run exceptionally fast without having to resort to anaerobic respiration. Alexander's argument boils down to the idea that there isn't necessarily an inherent physiological trade-off between sprinting and endurance but rather that there are costs to maximizing both that most other animals have never been forced to pay. Non-pronghorns instead focus on improving either one or the other, usually with the aid of specialized muscle fibers that pronghorns may or may not do without. Pronghorns do show at least one performance limitation as a result of their adaptations to long-distance sprinting, though: they prefer to avoid jumping for risk of breaking their slender legs, and when they do jump, they try to land hindlimb-first (which looks awkward and affords us a small degree of schadenfreude toward one of nature's most amazing athletes). This means that pronghorns are frequently stymied in their movements by ditches and trenches that other species can clear with ease.

What is the evidence that pronghorns have been selected for both sprinting and endurance in this way? If we look at the existing predators that bedevil pronghorns today, this idea seems shaky: the only terrestrial predators that pronghorns regularly deal with in the western prairies are coyotes, and they aren't going to be breaking any athletic records. In fact, the Road Runner in the classic Looney Tunes cartoons could well have been replaced with a pronghorn; Wile E. Coyote would have as little chance of catching one on the run as the other. But as pronghorn expert J. A. Byers points out in *Built for Speed: A Year in the Life of Pronghorn*, this paucity of predators is a relatively new development. Ten thousand years ago, during the Pleistocene, the western prairies (indeed, the North American continent) teemed with predators ranging from cheetahs, lions, and hyenas to fleet, long-limbed bears. It was during this (no doubt terrifying) time, Byers argues, that pronghorns faced sufficient danger from enough predators with varied performance capacities to evolve their unusual multipurpose morphology, overcoming or minimizing a ubiquitous functional constraint that other, less selectively motivated species never could.

Ironically, in doing so the pronghorns may have come to embody a different kind of evolutionary constraint—that of time lags, whereby the animals' functional capacities have not yet adjusted to their current ecological conditions both because the rate of response to relaxed selection is variable and can be slow and because ten thousand years is not very long in evolutionary terms. If this is the case, then the athletic abilities of pronghorns, amazing as they are, are overengineered for their current purpose and constitute nothing so much as relicts—the evolutionary legacy of a species haunted by the ghost of predation past.

Taking Strain

One way that animals can overcome the constraints inherent in muscle function or imposed via functional trade-offs is to do something entirely different. You've probably never noticed a froghopper in nature, nor is there any compelling reason why you should unless you are an entomologist or someone who really likes froghoppers. But there's a lot to like about these small and remarkable insects. Froghoppers are known in some parts of the world as spittlebugs because the nymph or immature stages of some types of froghoppers live in trees and cover themselves in foamy processed plant sap for protection, insulation, and moisture. That frothy substance sometimes drips down on unsuspecting passersby, creating the impression that certain trees are packed with little bugs just waiting for you to walk by so that they can spit on you, because nature hates you. But the most remarkable feature of froghoppers from a performance perspective is that they are the best jumpers on the planet, able to accelerate at almost four hundred times the force of gravity—roughly 75 percent of the average acceleration that is applied to a soccer ball when kicked by a professional player—and to leap more than one hundred times their own body length. Froghopper jumps even put fleas to shame because although fleas are also fantastic jumpers, froghoppers are on average sixty times heavier.

What makes froghoppers and fleas (and other insects such as grasshoppers, for that matter) such brilliant jumpers? Arnold's ecomorphological paradigm (see Chapter 6) tells us that the answer must lie in their morphology. For very small animals, jumping is a problem. Jump height is a function of the upward speed attained at the end of the acceleration phase that takes place while the animal is straightening its legs and pushing down on the ground, but small animals are limited in their ability to accelerate because of their short limbs.[3] This limitation comes about because animals can exert force on the ground to cause themselves to accelerate only as long as their limbs are in contact with it. But because the limbs of small animals are themselves necessarily small, they take very little time to straighten. Their limbs are therefore in contact with the ground only briefly during take-off, and their realized jump forces are correspondingly low. It would thus be sensible for small jumpy-type animals to evolve relatively longer limbs for their size, because they could then exert force on the ground for a greater length of time, and thus more force overall. We might then expect that froghoppers, fleas, and grasshoppers would have long limbs adapted for jumping. Grasshoppers do indeed have long hindlimbs, for exactly the reason I described, but the hindlimbs of fleas are relatively short, and frog-

3. In theory, all animals should be able to jump to more or less the same height in the absence of drag because the amount of work that an animal can do (force × distance) is proportional to its muscle mass. And since the percentage of an animal's mass that is made up of muscle is roughly constant in all animals, this means that work per unit mass is the same for all animals. The work done by an animal is therefore independent of the size of the animal. However, in practice this is not the case because of the limitations outlined above.

hopper limbs aren't notably long either. Limb length alone can't be the whole story, then, and something else must be going on.

Seahorse pivot feeding and the exceptional mantis and trap-jaw ant strikes described in Chapter 2 are enabled by elastic energy storage. A reasonable hypothesis is therefore that froghoppers and fleas have some comparable type of power amplification system based on stored elastic energy that trumps the requirement for lanky limbs. Malcolm Burrows of the University of Cambridge, who has spent much of the past twenty years studying aspects of jumping in invertebrates, shows that jumping froghoppers and fleas both use a catapult mechanism that combines slow-contracting muscle with a specialized elastic protein called resilin. Resilin has an elastic efficiency of almost 98 percent, which means that when it is stretched and then released, it gives back almost all of the energy that was used to stretch it, with only about 2 percent of that energy within it being lost as heat. This amazing property of resilin makes it ideal for power amplification, and froghoppers combine pads of resilin in each leg with a specialized bow-shaped part of the exoskeleton into a structure called the pleural arch that the animal flexes by contracting its jump muscles and cocking its legs. The cocking mechanism itself is also special: it contains the only known example in nature of a gear.

The froghoppers here take advantage of the aforementioned trade-off between muscle force and speed. By contracting their muscles very slowly, they maximize muscle contraction force and thus do more work to flex the pleural arches. Sudden relaxation of those muscles causes the arches to snap back into shape tremendously fast, with the recoil triggering the legs to extend rapidly, catapulting the animal forward at speeds and accelerations that could never be achieved with muscle alone. Froghoppers and fleas rely extensively on resilin to power their incredible jumps, but locusts and grasshoppers combine resilin catapult mechanisms in their knees with the mechanical advantage granted by their long limbs to power not only jumps but also kicks, should kicking another animal become necessary (which it often does). Resilin is also commonly found at the attachment points of insect wings to the exoskeleton, where it acts as a displacement amplifier, and is one of the reasons why many insects can flap their wings hundreds of times per second. What is more, resilin's elastic properties make it an ideal shock absorber, and in locusts it serves a secondary function of protecting the animal from self-harm if any of those powerful jumping or kicking motions go awry, forming part of a buckling region that absorbs energy during mishaps and reduces the force experienced by the rest of the animal.

Only invertebrates have resilin, but elastic storage mechanisms that contribute to performance are widespread throughout the animal kingdom. The alternative material that is commonly used for elastic energy storage is the protein collagen, which forms the majority (70–80 percent) of tendon tissue in vertebrates. The tensile strength of mammalian collagen (that is, the stress that causes it to break) is roughly two hundred times larger than the maximum stress that mammalian muscle can exert. This means that a very thin tendon can transmit the force of much thicker muscles without breaking, which is why tendons are used to attach muscle to bone. The extensibility of tendon—that is to say, the amount that tendon can be stretched beyond its normal length before failure—is remarkably low at 10 percent, and for its mass it stores elastic strain energy better than steel (but not quite as well as resilin). Put simply, tendon doesn't stretch very far, but it stores a large amount of energy when it does so, and like resilin, it gives back most of that energy when it is released.

Using an elastic element of low extensibility may seem odd, but it is important that tendons not be too stretchy, because they must transmit to the skeleton not only the force of the muscle shortening but the distance change as well. If you were to imagine a muscle attached to heavy bone by a stretchy rubber band, for example, you would find that contracting the muscle stretches the band so much that the connected bone moves only a little, if at all. At the same time, some extensibility is important—connecting muscle to bone with something that does not extend at all would transmit the distance change perfectly but preclude any elastic energy storage. Resilin (and to a lesser extent collagen) is the ideal biological compromise material.

So useful is collagen for storing and releasing elastic energy that it is found in vertebrate animals wherever powerful animal movements are required. Bush babies, for example, have a specialized muscle-tendon knee complex that allows these animals to store elastic energy in their leg extensor muscles and amplify power output by up to fifteen times, enabling them to achieve impressive vertical leaps. Collagen is also behind the ability of chameleons to shoot their tongues forward out of their mouths at accelerations of up to fifty times that of gravity—slightly larger than the maximum acceleration that a human has survived on a rocket sled.

Elastic energy storage mechanisms are useful for other reasons besides overcoming the limitations of muscle function. In salamanders that are often active at low core body temperatures, elastic mechanisms not only enhance tongue projection as they do in chameleons, but also buffer feeding behavior against variation in T_b. These salamanders have thus managed to liberate themselves from the tyranny of thermal performance curves by using elastic storage mechanisms to fire their tongues at similar speeds and accelerations over a T_b range from 2°C to 25°C. Although T_b still has some effect on projection performance (probably because muscle is required to load the elastic mechanism), that temperature effect is greatly reduced relative to other types of performance that do not rely on elastic energy storage.

Speed Limits

Animals like the cheetah, the pronghorn, and billfish may well live at or near the limits of what is possible in terms of their respective performance abilities by dint of the fact that they are the best. But could they be even better?

The question "What limits maximum performance?" has intrigued researchers for years, and as is often the case, there isn't one simple answer. In addition to the constraints placed on a particular type of performance by the need to perform other, potentially conflicting, athletic feats, performance limits comprise a combination of factors of varying relevance to different species, some of which delve into increasingly esoteric realms of physiology. Rather than risk alienating those of you who aren't my mother and picked this book up only to learn cool stuff about animals, I'll focus on a few of the more general ones.

A popular suggestion for a factor that limits maximum speed is the capacity of muscle to produce power (which is, again, the ability to apply at a high rate force that results in movement). Individuals that produce high mechanical power should be able to attain high speeds. In the case of flight, the relationship between power and speed is complex, but it is nonetheless probably safe to say that power output is a limiting factor on flight capacities, and also on other types of performance abilities that rely on short bursts of high power, including jumping.

For other types of performance, however, it isn't always apparent whether individuals are producing the absolute maximum power they are capable of. To test the hypothesis that power output limits sprint speed, for example, we would need to place our test organisms under conditions that elicit increased power output and see whether the speeds are maintained compared to control conditions. If they are not, then power is likely a limiting factor on speed. In the case of sprinting specifically, one way to do this is to compel an animal to sprint at its maximum speed up steeper and steeper inclines, which demand more and more power because the animal has to work increasingly against the effects of gravity as incline increases. If the animals can perform just as well on inclines as on flat surfaces, then the hypothesis of power limiting performance is falsified.

Reptiles are especially adept at generating high power, likely because of the differences between ectotherm and endotherm muscle that I mentioned in Chapter 5. Just as with gravid female green iguanas, maximum speeds appear not to be limited by power output in those lizard species that have been tested in this way. The western banded gecko and the western skink each generate concomitantly higher power outputs as they run up inclines varying in steepness from zero to 40 degrees, resulting in similar speeds at all inclines. This suggests that lizards running along a horizontal surface at their maximum speed have power to spare. Another way to get at this question is to increase the mass of animals that are already climbing vertically, again requiring their muscles to output more power to move the increasing loads. If you were to go rock climbing with two other people of the same weight as yourself hanging off your back, I would make the very safe prediction that your climbing speed would decrease dramatically as your muscles fail to produce sufficient power to maintain your unloaded speed. Duncan Irschick of the University of Massachusetts at Amherst and his collaborators found that in both Tokay and Mediterranean house geckos, this isn't what happens: both species increase their power output as required to match their unloaded maximum climbing speeds, even when weighed down with masses up to 200 percent of their total body mass!

Yet another lizardcentric approach has been to manipulate mechanical power output by changing body temperature, and studies that have done so have found that here power output may indeed limit sprint speed—but only below 25°C (77°F), and not at core body temperatures close to the animals' optimal sprinting ranges. Data are still scant, but it appears for now that power output does not limit sprint speed. Instead, biomechanical factors such as the amount of time an animal's feet are in contact with the ground and the elastic properties of the musculoskeletal system (as touched on in the gravid green iguanas of Chapter 4) may be more important for terrestrial movement.

Power output may also contribute to speed limits in aquatic organisms, but in a very different way. The eight hundred times higher density of water relative to air has important consequences for animals that move through water very quickly at high Reynolds numbers, especially for animals such as fishes and cetaceans (that is, dolphins, porpoises, and whales) that swim via rapid oscillations of a crescent-shaped tail fin. That fin acts in similar fashion to an airfoil, generating lift (laterally in the cases of fishes with vertically oriented tail fins, dorsally and ventrally in cetaceans with horizontal fins)[4] that drives the animal forward just as flapping wings do.

At high enough speeds, the pressure of fluid at the front (leading) edge of that tail fin drops below the vapor pressure of the fluid, causing the formation of vapor-filled cavities. In other words, bubbles start to form. Those bubbles are then carried downstream, back over the fin, to regions of higher pressure, where they collapse, rapidly releasing energy (you will recall this phenomenon of cavitation from Chapter 2, when I discussed how mantis shrimp harness it to capture prey). These collapsing bubbles can damage

4. The difference in orientation between the caudal fins of fish and cetacean flukes are the result of cetaceans and fish converging on a similar method of swimming from different ancestral starting points. Ichthyosaurs—extinct marine reptiles—also exhibited the vertical fin orientation.

the surface of the swimming appendage itself, but even if they don't and only collapse farther downstream of the back (trailing) edge of the fin, they can disrupt fluid flow and cause a cavitation-induced stall that slows the movement of the entire animal through water. Instead of aiming to produce sufficient power to move the center of mass, as is required for flying or jumping, fishes and cetaceans need to be careful that they don't produce too much power, which would expose them to either cavitation damage or a performance decrement through cavitation-induced stalling. These physical limitations of the aquatic medium therefore serve as speed limits on animals that move in this way, and those limits vary based on factors such as the size of the animal and the depth at which it is moving, with large, deep swimmers being more at risk from stalls and large animals at the surface facing increased hazard from cavitation damage.

The above scenario is proposed in a hydrodynamic model of fish swimming generated by engineers at the Technion in Israel, but it is based on empirical swimming data in actual animals. Like all models it should be independently verified, but additional evidence lends support to its predictions, such as the observation that cavitation-induced damage has been detected on the fins of scombrid fishes (such as tuna and mackerel), which lack pain receptors in those fins, whereas dolphins, whose fins certainly do feel pain, are never observed to approach their theoretical speed limits. But even for fish that are unlikely to move fast enough to be slowed by the environment power remains a key constraint, and some teleost fish are limited in their fast-start (that is, acceleration from a standstill) performance by muscle power output in ways that terrestrial animals such as lizards are not.

It's All in Your Head (?)

Power output may be a limiting factor on speed and on burst performance in general, but it does not place similar constraints on endurance. The study of endurance in humans is light-years progressed past what is known in other animals, and so it is logical to consider human endurance limits here.

The traditional cardiovascular model of human exercise physiology that has held sway since the 1920s says that endurance abilities are not only enabled by the body's ability to supply oxygen (as I have emphasized repeatedly) but also limited by it. Under this model, the extreme fatigue experienced by competitors toward the end of marathons and ultramarathons is a symptom of the body's decreasing ability to supply sufficient oxygen to the muscles and other parts of the body that need it, and exercise is ultimately terminated once the limit of that ability is reached and fatigue becomes overwhelming. This paradigm has since been challenged. Critics of the cardiovascular model suggest that if it were true, then the first victim of inadequate oxygen supply during endurance exercise would be the organ that requires oxygen the most—namely, the heart muscle itself. This in turn means that cessation of exercise should be driven by extreme cramping accompanied by cardiovascular failure, but since it is apparently not usual for the final stretches of marathons to be littered with the bodies of athletes in various stages of cardiac arrest, something else must be happening.

The argument hinges partly on the nature of fatigue and what causes it. Fatigue is a complicated and multifaceted phenomenon, and although lactate build-up is the widespread (and probably fallacious)[5] explanation for muscle fatigue in particular, the reality is more complex. In fact, exercise physiologists don't have a good handle at all on the physiological causes of fatigue, which has led critics of the cardiovascular model, foremost among them Tim Noakes of the Sports Science Institute at the University of Cape Town, to suggest an alternative.

Noakes's explanation is an updated version of an idea originally proposed in 1924 by the Nobel Prize–winning physiologist A. V. Hill. According to Noakes and Hill there lurks in the recesses of the human brain a region called the central governor that is tasked with overseeing and regulating endurance exercise. The job of this hypothetical brain region is to ensure

5. Lactate, frequently thought of as merely a byproduct of anaerobic respiration, is an enormously misunderstood molecule that is deserving of more space than I have to dedicate to it here. Suffice it to say that lactate likely does not cause fatigue and is in fact an important supplemental fuel (especially in ectothermic vertebrates) that is shuttled around the body during exercise and is produced in small amounts by muscle even when oxygen delivery is perfectly adequate. Experiments also show that lactate supplementation has no effect on muscle soreness or fatigue in humans.

that the actual limits of cardiovascular function are never reached, and thus to enforce the termination of exercise by limiting skeletal muscle activity before the heart itself becomes damaged. It does this by inducing a perception of fatigue in the exercising animal. This area of the brain would have to receive neural sensory input regarding body temperature, limb positions, oxygen and carbon dioxide levels in the blood, and cardiac function. All of these sensory feedback mechanisms do in fact exist, and we also know that there are several situations where the brain acts on this sensory feedback to alter aspects of the cardiovascular circulation. The mammalian dive reflex, for example, slows an individual's breathing and heart rate when the face comes into contact with water. So under the central governor hypothesis, fatigue isn't a cumulative physiological phenomenon driven by increasingly inadequate oxygen supply. Instead, it is a sensation, like pain, that is experienced only by the brain even though it appears to originate elsewhere, and it exists to prevent us from overexerting and damaging ourselves.

Opponents of the central governor (of which there are many) are of course aware of these neural mechanisms, but they argue that it isn't necessary to conceive of a region of the brain that integrates them all because they clearly work well enough without one. A further objection is that the limits of cardiac function, respiration rate, and oxygen movement are already set by physical laws, not by any neural regulators. The biggest strike against the central governor, though, is that no one has found any evidence that such a region of the brain exists. Sticking electrodes inside living human brains is now frowned on, which means that the only way to establish the central governor's existence would be to place exercising humans inside MRI machines and monitor their brains for regions of increased activity as fatigue intensifies. Since such machines are not built to accommodate treadmills, nor are they likely to be, that's a potentially insurmountable logistical problem.

The central governor model is extremely contentious. A lot of sports physiologists *really* don't like it. Proponents for and against have exchanged opinions, commentaries, and critiques in the literature that range from reasonable criticisms of each other's positions to scathing personal attacks exhorting specific individuals "and those of his ilk" on each side to "put up or shut up." I have no investment in this hypothesis either way other than to re-mark that some dodgy appeals to evolution have sometimes been deployed, and proper mud-slinging in the scientific literature is rare enough that I find these vitriolic exchanges highly entertaining. But they resolve little. To my mind, the most salient points are, first, that neural control mechanisms for feedback of exercise to and from the brain do exist, so a discrete region of the brain that integrates them isn't a totally wacky idea, and second, that there is no evidence, and possibly no need, for such a region.

Rather than speculating as to the selection pressures that might work for or against a possibly imaginary central governor, as some have done, it seems reasonable that if such a region does exist in humans, then it might also be present in some other likely candidate animals that are more ame-nable to exercising inside an MRI machine. However, that suggestion is already out there and it did not go down well.

It is worth noting, though, that neural mechanisms involved in exercise are of increasing interest to biologists, and some fascinating findings are coming to light. For example, a particular region of the brain called the me-dial entorhinal cortex integrates memory and navigation. In 1995, a group of researchers including John O'Keefe, May-Britt Moser, and Edvard Moser showed that rats encode a neural map of their spatial environment into that cortex—a finding that won them a Nobel Prize in 2014. In 2015, that same research group showed that the entorhinal cortex also harbors a group of neurons called speed cells that increase their activity in direct proportion to the speed at which the animal is moving, regardless of the context of move-ment. Thus, rats measure how fast they are running and combine this in the medial entorhinal cortex with their neural map of the environment to figure out exactly where they are in that environment and how rapidly they are moving through it.

Even though it might seem obvious that animals would be monitoring both their own speed and position in the immediate environment (to re-member the location of nearby obstacles, for example), these studies are vital for understanding the neural mechanisms by which this happens. The temptation to speculate that the entorhinal cortex might constitute or be af-filiated with a central governor here is strong, but no one has suggested that the governor's purview might extend to speed (nor is there any compelling reason why it should), and in any case the conceptual leap from speed-sensing cells to a central exercise governor is large, to say the least.

Ecological and Optimal Performance

The hypothesized existence of a central exercise governor may seem like an amusing and ultimately untestable thought experiment, but it pertains directly to an important and often neglected question as to how often animals really do approach their performance limits in nature. Indeed, there is a good reason why thus far I have concentrated almost exclusively on the maximum capabilities of animal performance that are usually measured in the laboratory and have said relatively little about *ecological performance*— that is, the extent of their maximum performance abilities that animals use in the wild when left to their own devices, unmolested by prying scientists and unperturbed by artificial environments.

Performance researchers have turned their attention to this issue only occasionally, and there are several reasons why we have thus far failed to build up a substantial literature on ecological performance. The first is that obtaining data on the type and extent of performance abilities that animals use in nature has historically been very difficult and, for some types of animals, bordering on impossible. Fortunately, advances in technology, from miniature GPS tags and accelerometers to GoPros and portable high-speed cameras, mean that collecting data from animals in motion in the wild is now easier than ever.

But a perhaps subtler impediment to our understanding of ecological performance is a particular quirk of the way science sometimes operates. Originally researchers *were* primarily interested in measuring, for example, the speeds at which animals move in nature but, lacking the means at the time to measure it with precision, instead devised methods for measuring maximum performance under controlled laboratory conditions. Because these methods proved to be so successful, subsequent researchers continued to measure maximum laboratory performance almost exclusively. Consequently, the original intention of understanding ecological performance in nature fell by the wayside, and over time the idea that we were measuring something meaningful—incidentally embedding in the field the implicit assumption that maximum and ecological performance are equivalent—went largely unquestioned.

The comparatively few studies that have tested the assumption of equivalency between maximum and ecological performance have shown that animals almost never use their maximum performance abilities in nature! For example, despite all the ink that has been spilled over cheetahs' speed, the studies of free-ranging cheetahs I mentioned earlier show that cheetahs never reached their maximum speed during the recorded hunts and that the distribution of speeds attained during successful hunts does not differ from that observed for unsuccessful hunts. Similar results have been reported for various species of lizards as well as several other animal species. These data on ecological performance raise a number of questions, including why and under what conditions animals modulate their performance abilities and, most interestingly from an evolutionary perspective, why animals have evolved specialized and expensive performance capacities that they seldom take full advantage of.

Potential evolutionary explanations to this question, such as time lags in the case of pronghorns, have received little attention and would in most cases be difficult to test anyway. A particularly interesting idea regarding why there is variation in the extent to which animals use their maximum performance capacities has been proposed by David Carrier of the University of Utah. Carrier argues that juvenile animals use a greater proportion of their maximum performance capacities than do adults of the same species because juveniles are more vulnerable due to their smaller size. It thus behooves them to go about their business especially quickly before something kills and eats them. This selection on juvenile performance could have carryover effects that persist into adulthood, resulting in adults with greater maximum performance capacities than they actually need.

The core component of Carrier's idea, called the *compensation hypothesis* (because juveniles perform at high levels to compensate for being small), is supported in several animal species. For example, juvenile collared lizards do indeed use a greater percentage of their maximum speed in nature than adults do, and hatchling earthworms displace more earth for their size than adults when burrowing, even after accounting for scaling effects.

Explanations other than the compensation hypothesis for the variation seen in ecological performance are thin on the ground. To be clear, though, this mismatch between maximum and ecological performance doesn't necessarily mean that maximum capacities are not important, and we have good reason to believe that selection has acted on maximum performance capacities. For example, minnows with both the fastest swimming speeds and the highest swimming endurance are the least vulnerable to being captured through trawling, suggesting that there has been strong selection for maximum swimming ability (even though that selection has been exerted by human beings).

It may well be that there are many other cases where maximum performance abilities serve as a buffer against rare or infrequent selective events such as predation attempts that have a disproportionate effect on fitness, and it thus serves animals well to maintain those high-performance capacities even though they are seldom used. Indeed, this very result emerged from a simulation model of performance evolution built by my former graduate student Ann Cespedes as part of her dissertation. Furthermore, a study by Jakob Bro-Jørgensen of the University of Liverpool found that maximum sprint speeds of African savanna herbivores are predicted by vulnerability to their major predators, as would be expected if speed were an important part of escaping predation in these animals. But even if we can tell a lot about an animal's ecology and fitness prospects from studying those maximum abilities, we nonetheless have a long way to go toward understanding how animals use those capacities in the field and about the various factors that affect performance evolution.

Recent thinking on (especially) the speeds that animals use in nature has focused on the balance between the benefits of moving fast and the potential costs of doing so. Animals should strive to minimize those costs while simultaneously maximizing the benefits, resulting in characteristic *optimum speeds* for each species. Although costs to performance can come in various forms, biologists have long focused on energetic costs in particular. Measuring those costs can be a challenge, but a great deal of evidence suggests that making use of maximum performance abilities all the time would be energetically very, very expensive.

EIGHT

Death and Taxes

B eing alive costs energy. Doing things while being alive, be they locating a mate or finding, catching, and even digesting food, costs even more. The requirement for energy to fuel both being alive and activities beyond merely existing is why animals need food in the first place, and the amount (and sometimes the type) of energetic resources that animals can extract in the form of food from their environment influences the kinds of performance abilities that those animals can support. This simple fact is the basis for a multibillion-dollar sports nutrition industry in humans, but it also has real implications for animals in nature. Performance isn't free, and because few animals live in environments where food is unlimited, the amount of energy an animal spends on performance also affects (and sometimes limits) how much it can spend on other important energy-consuming processes. The economics of energetic acquisition and expenditure thus has clear consequences for individual fitness. And because energy budgets and priorities for spending that energy are dynamic and may change depending on the life stage of an individual, they can influence not only how that animal lives its life and how it ages but how long it will likely live as well.

The Economics of Performance

The rate at which animals spend energy is a fundamental concept in evolutionary physiology. We can measure the rate of energetic expenditure (that is, the metabolic rate) in several ways, but the most popular is to measure the rate of oxygen consumption. This works because aerobic respiration provides the energy that fuels almost all basic physiological processes in animals.

Aerobic respiration takes the basic components that the process of digestion breaks down foods into—carbohydrates, fats, and protein—and converts them, with the aid of oxygen, into a form of energy currency called adenosine triphosphate (ATP) that can then be used to meet the energetic demands of cells, tissues, and organs. We can use the rate at which an animal consumes oxygen as a proxy for its metabolic rate because a liter of oxygen used in aerobic metabolism produces roughly the same amount of usable ATP regardless of whether fats, protein, or carbohydrates are being oxidized. This turns out to be convenient for physiologists, who can directly compare metabolic rates across different kinds of animals regardless of those animals' diets. It's also great for performance researchers, because it means that we can measure the energetic costs of all kinds of aerobic activities, so long as we can convince the animals to undertake those activities while inside or attached to an experimental setup that measures the amount of oxygen they consume while doing so.[1]

The energetic costs of activity depend on numerous factors ranging from the type of performance involved to the size of the animal in question. For locomotor performance in particular, the medium through which the animal is moving and the speed at which it is doing so strongly influence the energetic costs of performance. Of the three primary modes of locomotion—swimming, flying, and moving on land—swimming is the cheapest, flying the next cheapest, and terrestrial locomotion the most expensive.

The reason why swimming is relatively cheap is because the density of water goes a long way to supporting the body weight of any animal that finds itself on or under water. A swimming animal therefore has to do little work to counteract the influence of gravity. To be completely neutral, the density of the animal must be equal to the density of the medium surrounding it (the ratio of the two is the *specific gravity*). Specific gravities of aquatic animals are commonly close to, but slightly larger than, 1, which means

---

1. An unattributed maxim known as the Harvard Rule of Animal Behavior states that "under perfectly controlled conditions of light, temperature, and humidity, the organism will do whatever it damn well pleases."

that they can float or sink only very slowly. Some animals have adaptations that further reduce specific gravity, such as gas-filled swim bladders, or the lighter skeletons found in sharks and rays that are made out of cartilage as opposed to bone.

For flying animals, the ability to use lift to overcome gravity is handy in terms of mitigating the energetic costs of flight. Measurements of those costs in the laboratory show that there is a U-shaped relation between flight speed and metabolism. This means that it is expensive to fly very slowly or very fast and cheapest to fly at some intermediate speed. Hovering (that is, flying at zero speed) is particularly expensive. Measurements of energetic expenditures during hovering in hummingbirds show that metabolic rates during hovering are ten to twelve times the *basal metabolic rate*. This means that hummingbirds spend energy ten to twelve times faster during hovering than when doing nothing at all, not even digesting food. The nectar-feeding bat *Glossophaga soricina* similarly expends energy at twelve times its basal rate when hovering. Variation in flight morphology among hummingbirds contributes to variation in this regard; for example, territorial species have shorter wings that contribute to maneuverability but create higher wing loadings, resulting in greater costs of hovering.

Terrestrial locomotion presents very different problems to either swimming or flying. Land animals have to use their legs as levers against the ground while supporting their centers of mass above said ground at some optimal level, and they do energy-consuming work to both accelerate and decelerate that mass. Terrestrial animals even expend energy just to stay upright! Neither of these things are easily achieved, and both contribute to the cost and relative inefficiency of walking and running. Legged locomotion is so inefficient, in fact, that we humans can achieve much better results by attaching ourselves to bicycles and powering them with our muscles instead. The energy savings from doing so are far from trivial, and at a speed of 2 m/s (7.2 kph, or 4.47 mph), it costs 2.2 times less to cycle a given distance than it would to walk that same distance at the same speed. At twice that speed, the savings are 3.7 times.

Last, the costs of performance are strongly influenced by the size of the animal regardless of whether that animal is swimming, flying, or running.

A variety of things, from muscle function to metabolic rate, are relatively more expensive for smaller animals, at least partly because of the scaling relations between area and volume that I mentioned in Chapter 6. Small animals thus face significant energetic challenges. For example, it takes up to a month for each gram of elephant tissue to consume as much oxygen as one gram of shrew tissue uses in one day when both animals are inactive!

Elephants are of course made up of many more grams of tissue than shrews and thus use more energy per day overall, but small animals spend more energy *per hour per unit size* than do larger ones. This means that small animals burn through whatever energy stores they might have very quickly. The relative energetic requirements of small hummingbirds, for example, are so exorbitant that they must feed almost constantly and are forced to lower their basal metabolic rates significantly overnight to avoid starving to death in their sleep. For much the same reasons, the relative costs of performance are also higher for smaller animals than for larger ones. This also explains a number of other general differences in lifestyle between small and large animals. Few small animals migrate seasonally, for instance, not because they cannot cover large distances quickly enough, but because they cannot store enough energy to pay the relatively more expensive costs of doing so.

No Free Lunch

Because the costs of performance can be influenced by all of the above factors, it is difficult to put a number on exactly how costly performance is in an animal's daily life in the wild. Animals that regularly use the full extent of their locomotor abilities would probably incur very large energetic expenditures, but as we saw in the last chapter, this is not likely. In 1983, Ted Garland of the University of California, Riverside, developed a metric called the ecological cost of transport (ECT) to estimate what percentage of an animal's daily energetic expenditure is accounted for by the costs of locomotion in particular. To derive this metric, we need to know three things: the total distance an animal moves every day, the energetic cost of moving that distance, and the amount of energy that animal spends per day in nature[2] over and above its baseline metabolic costs.

For the vast majority of animals we know none of these things, and for a few, only some. At the time of writing, we can calculate the ECT in this way for about seventy species of mammals. The ECTs of mammals show enormous variation, with some mammals devoting only a few percent of their daily energy budgets to support locomotor activities. Carnivores show the highest ECTs, with some carnivorous species spending more than 20 percent of their daily energy expenditure on locomotion alone, very likely driven by the longer distances that many carnivores move in search of food. Of course, the ECT is only an estimate based on snapshots of animal activity, and we cannot know for sure what animals are up to without following specific individuals closely.

A study led by Terrie Williams of the University of California, Santa Cruz, used an innovative combination of traditional metabolic rate measurement and SMART (Species Movement, Acceleration, and Radio-Tracking) collars, similar to those used to measure movement in free-ranging cheetahs, to assess the energetic costs of ambush hunting in pumas with more precision than ever. The amount and type of data that the researchers were able to glean from the SMART collars are a functional biologist's wet dream: Williams and colleagues estimated speed, acceleration, cost of locomotion, time spent moving, type of activity (hunting versus nonhunting), type of hunting (chasing versus stalking and pouncing), how often the animal turned, what type of terrain it was moving on, whether prey capture occurred, and even the size of the captured prey! What is more, by measuring the energetic costs of various activities in pumas in the laboratory and creating a library of movement "signatures" that they were able to associate

2. We can estimate these costs by injecting free-ranging individuals with known amounts of a special kind of water called *doubly labeled water* made up of isotopes (that is, heavier than normal versions) of hydrogen and oxygen. From blood samples collected at a known later date, we can measure how much of the labeled oxygen remains in the system and thus how much oxygen was consumed by the animal. Tracking hydrogen as well lets us correct for oxygen lost as water, for instance in urine.

with those measured costs, the researchers calculated the costs incurred by free-ranging pumas equipped with SMART collars based on the occurrence of those movement signatures as the animals were going about their daily lives. (Once again, this innovative and important study resulted from one of the treadmill-based NSF grants that Senator Coburn decried as wasteful and pointless.)

In this way, the researchers worked out not only where the animals were and what they were doing at any given time but also how much energy they were expending doing so. They were therefore able to show, for example, that the energetic costs of locating prey (what they called the "pre-kill hunting costs") accounted for 10 percent to 20 percent of their total energy costs, whereas the costs of prey capture itself varied depending on the size of the prey. They thus concluded that the evolution of ambush hunting in these animals was mostly driven by the requirement to minimize the energetic costs involved in locating and subduing prey (which, of course, involves performance). This study puts exact numbers on the costs of using locomotor abilities in nature and shows that they really do constitute a major energetic expense. In fact, costs of performance can be so profound that animals go to great lengths to make such activities as they are required to undertake as efficient as possible.

Saving Energy

Animals minimize the energetic costs of performance in different ways. The European kestrel often hovers when searching for prey, but only when flying directly into the wind, which is energetically equivalent to flying horizontally at that particular wind velocity. Kestrels hover most frequently at wind speeds that correspond with the cheapest cost of flight and less frequently at wind speeds that are faster or slower than energetically optimal, which strongly suggests that these birds are attempting to lessen the energetic costs of hovering.

European shags (a type of cormorant) use this same principle by preferentially taking off directly into headwinds. Just as in kestrels, the wind moving over their wings at speed generates lift without the animal having to do so itself through beating its wings, resulting in significant energetic savings and allowing these birds to get away without maintaining the large and expensive flight muscles that other birds use primarily for take-off—on average, species of modern birds that devote less than 16 percent of their body mass to flight muscle have at best marginal ability to take off under their own power. Soaring and gliding are also useful for saving energy while traversing relatively long distances, and some birds, bats, and large insects such as butterflies further maximize flight time and/or distance by gliding close to water or ground surfaces, which results in an upwash that reduces drag and improves lift, a phenomenon called *ground effect*. Somewhat bizarrely, great hammerhead sharks are reported to spend 90 percent of their time swimming on their sides, at angles between 50 and 75 degrees. At these orientations, the animals' extremely large dorsal fins contribute significantly to lift, reducing energetic costs of hammerhead transport by 10 percent.

Penguins, as already noted, employ an amusing waddling gait when on land that does them no favors in terms of conserving energy and is a result of their compromise morphology for swimming. Because waddling is so inefficient, they use wherever possible a far cheaper behavior called tobogganing, which involves them lying on their bellies and using their flippers to push themselves along on snow, especially down snowy slopes. Tobogganing is not the only alternate means of locomotion that penguins use; when swimming near the surface, penguins also leap repeatedly out of the water, a behavior called porpoising (because porpoises also do this, and because "penguining" almost certainly already refers to some kind of fetish).

On the face of it this seems like a terrible idea. Jumping out of the water over and over again seems expensive, but mechanical analyses suggest that it is actually an energy-saving means of locomotion. Swimming close to the surface is harder than swimming deeper underwater because surface swimming generates waves. When this happens, some of the energy of the moving animal is lost as it is converted into work done forming and moving waves, ultimately slowing the animal, a phenomenon referred to as *wave drag*. For penguins specifically, swimming normally at the surface requires 50 percent more energy than swimming underwater at the same speed, primarily as a result of this additional wave drag. But if the animal is moving fast enough at the surface (that is, above threshold crossover speeds that work out to be around 11 kph/6.8 mph for Humboldt penguins and 18 kph/11.2 mph for a much larger dolphin), then the reduced drag experienced during the aerial phase of porpoising while the animal is traveling clear of the water more than offsets the additional cost of leaping, resulting in a net energetic savings.

Although this seems like the kind of counterintuitive and spectacular adaptation that is tailor-made for people writing books about performance, the mechanics involved in porpoising are, unfortunately, complex, and the interpretation of this behavior is thus controversial. So whereas some estimates suggest that dolphins that alternate burst swimming-powered porpoising with extended periods of underwater coasting could realize energetic savings of up to 40 percent compared with swimming only, other biologists are skeptical and note instead that only air-breathing animals porpoise and that perhaps this fact has something to do with it. Whether or not porpoising is effective in saving energy, dolphins already swim more cheaply than other aquatic mammals, and bottlenose dolphins save yet more energy by wave-riding (or body-surfing) on surface waves.

Animals that need to supply power above and beyond what their muscles can generate for burst activities may resort to elastic energy storage (as discussed in Chapter 7) to make up the shortfall. But another common use of these elastic systems is to reduce the amount of muscular work involved in sustained locomotion, thereby making locomotion cheaper. Terrestrial locomotion in particular requires limbs to be sequentially accelerated and decelerated over and over, and so if there is a way to store the kinetic energy that is lost in the first part of a stride and use that energy in the next part, then the overall energetic cost of the stride is diminished.

Tendons, and to a lesser extent muscles, do this, and because the tendons in particular store and give back energy so efficiently, they can increase the economy of motion by recycling energy. Kangaroos spend energy getting up to speed but can hop cheaply at high speeds thanks to their enormous Achilles tendon, which acts like a spring, storing and releasing energy as the animal transitions into bipedal hopping above a certain threshold speed. In effect, the animal bounces when hopping as if on a pogo stick, which is a remarkably efficient way to travel at high speeds, reducing the amount of work required of the muscles by up to 45 percent in some types of wallabies that do the same thing.

Many animals, from camels, horses, and antelopes to monkeys and even humans, use elastic energy storage to reduce the costs of locomotion. Perhaps the most effective are turkeys, which can save up to 60 percent of the costs of running through elastic energy storage when running at high speeds. But there are also some animals to whom these savings are unavailable; compromise morphologies in particular, such as those between swimming and running, are probably less efficient at storing and recovering energy than those specialized for one locomotor mode, and large animals are also likely to benefit more from elastic energy storage than small ones through their higher mass.

Friends with Energetic Benefits

Almost everyone has seen, at some time or another, birds flying in formation. The V-shaped formation adopted by large migratory birds such as ducks and geese, for example, is so commonplace that few give a second thought as to why these birds adopt the V shape in particular, as opposed to some other configuration. Those reasons likely have to do with the energetic savings gleaned from performing in groups. A beating wing produces a vortex above the wingtip that rotates in the vertical plane as the wing changes direction from upstroke to downstroke. Mosquitoes rely extensively on these vortices for flight, as mentioned in Chapter 6, but birds that fly in V-formations save energy by deriving lift from the vortices produced by the wingbeats of birds preceding them, similar in concept to slipstreaming in cycling. This requires some overlap of wingtips to take full advantage, hence the characteristic V configuration as birds arrange themselves so as to coast on the vortices of their neighbors. Pink-footed geese may save only 2.5 percent of the total flight cost by doing this, whereas greylag geese save between 4 percent and 9 percent of the costs of flight in this manner. That may not sound like much, but over the long distances that these migratory animals fly (from nesting sites in Greenland and Iceland to winter in northwestern Europe in the case of pink-footed geese), those savings can add up. These potential energetic savings also explain why V-formations are found only in relatively large birds that weigh more than a few kilograms; small species probably wouldn't benefit in this way due to both the small vortices that their wings generate and the general lack of wingtip overlap among their shorter wings.

The energetic benefits of doing things in groups extend to other forms of performance besides flight. Sea turtles, although solitary as adults, are born into large groups of hatchlings because female sea turtles lay clutches of between 50 and 150 eggs into underground nests that she digs into the beach herself and then covers up with sand. All eggs within a clutch therefore share the same nest and experience a similar incubation temperature, which, thanks to turtle hatchlings' ectothermic nature, means that they all grow and develop at more or less the same rate and consequently all hatch at the same time.

Synchronous hatching is an important feature of sea turtle reproduction, to the point that some have suggested that hatchlings in some species can stimulate the hatching of tardy clutch mates that have yet to begin the hatching process. But before you start Googling pictures of adorable baby sea turtles and posting to social media about how heartwarming it is that they make a point of all entering into the world together (#NoBabySeaTurtleLeftBehind), consider that animal group activities are typically undertaken only because participating in that group affords each participant some kind of benefit over going it alone. In this case, that benefit is that it is cheaper and easier for heaps of baby sea turtles to dig themselves out of an underground nest at the same time than it is for a maverick baby sea turtle to tunnel out alone.

Larger cohorts of hatchlings show both shorter digging times and a lower energetic digging cost per individual when digging upward through 40 cm (15.7 in.) of beach sand, with an increase in group size from ten to sixty animals being equivalent to a 50 percent reduction in digging duration and cost. Conserving energy from the very start is crucial for sea turtle hatchlings because they head immediately for the ocean from their hatching sites on the beach, running a gauntlet of predators waiting for the baby sea turtle buffet to emerge. Any given baby turtle stands a greater chance of survival if there are lots of other baby sea turtles to share the risk of predation with— yet another potential benefit to emerging in groups. Those lucky hatchlings that make it to the ocean, where their small size means that they are buffeted by the pounding surf, then face a continuous twenty-four-hour swim, called a swimming frenzy, with the aim of escaping the predator-laden nearshore waters and reaching the (relatively) safer open waters as soon as they possibly can. The swimming frenzy is a literal race for the baby sea turtles' lives. With some estimates of hatchling survival to adulthood as low as one in one thousand, these little animals need all the help they can get.

The substantial energetic incentives derived from group living are one of several factors driving the evolution of sociality in several animal species, with at least some of those incentives pertaining specifically to performance. Sociality is rare among vertebrates but has nonetheless evolved several times despite a variety of associated costs such as increased rates of parasite transmission among individuals that find themselves near to one another. For some species, the benefits to sociality include reduced energetic costs of foraging and/or prey capture compared with a solitary lifestyle. As noted above, carnivorous lifestyles are energetically expensive, particularly for mammals. Around 7 percent of vertebrate carnivores exhibit some form of cooperative hunting, and although those species vary in their hunting strategies, cooperative hunting enables them to share the energetic burden of prey capture, often allowing them to capture and subdue more or larger prey than they could take on alone. Pack members of domestic dogs take turns harrying prey individuals over distances as long as 19 km (almost 12 mi.) at average speeds of roughly 6.5 kph (4 mph), driving the prey to exhaustion faster than any individual dog could alone, and GPS data on wild dog hunts show that the benefits of such group hunts in terms of hunting success and feeding frequency greatly outweigh the energetic costs of repeated short chases. Even some invertebrates, such as social spiders, cooperate to capture prey larger than any individual spider can handle, presumably resulting in a larger energetic payoff than solitary hunting.

Trading Off Performance and Life History

It is not only performance that costs energy to fuel and use. Animals must operate, often concurrently, several physiological processes that each account for significant chunks of their overall energetic budget. To give but one example, one of the most expensive things an animal might do over the course of its lifetime is reproduce, and many animal species will do so more than once. The female cotton rat, for example, increases her energetic expenditure by 37 percent over the period from pregnancy (twenty-six days) to weaning of her five young (the subsequent twelve days). But those costs are not distributed equally throughout the reproductive period; the vast majority of them occur over the twelve-day weaning period, when the mother is producing milk (or lactating) for her pups. Lactation is especially pricey, and female cotton rats spend energy five times faster during lactation compared to the rest of pregnancy.

Reproduction in other nonmarsupial mammals increases energetic expenditure overall by around 25 percent on average, with about 80 percent of that additional expenditure accounted for by lactation alone. These additional costs are also not limited to mammals. Birds do not lactate but nonetheless face similar energetic costs to provisioning their offspring and increase their metabolic expenditures threefold when feeding their young.

Any energetic expenditure must factor into an animal's energy budget, but it should be clear that some costs incur a larger energetic burden than others. It is important to differentiate between things that merely cost energy insofar as they require some amount of energetic expenditure and those that are expensive and account for a large proportion of an animal's overall energetic budget. If we were to add up all of the potential energetic costs that individuals of a species are subject to at any given time — including the costs of expressing certain traits or signals, costs involved in finding food or mates, costs involved in display or combat, and costs involved in reproduction, to name only a few, and some of which are more costly than others — we would find that the total amount of energy required to support all of them at the same time is very large, and certainly far larger than the amount of energetic resources that an individual has at its disposal at any given time. This means that individuals with limited energy budgets cannot invest maximally in everything at the same time and must therefore prioritize which traits to invest those resources in, and which traits not to invest in, just as people with limited monthly financial budgets have to decide what things they can afford to spend money on that month and what things they can't.

This analogy works because once you have spent money on something, that money is gone and you can't spend it again on something else (unless you work on Wall Street). So it is for energy. The amount of energy that an organism can afford to invest in a particular trait therefore depends on the amount of energetic resources (that is, food) that it has acquired from the environment as well as the energetic price of that trait. Unless those resources are unlimited, which (spoiler alert) they almost never are, the energetic investment decisions the organism makes, and the relative costs of investing in particular traits, lead to trade-offs in trait expression. Monetary trade-offs in humans mean investing first in essentials like groceries and shelter, which leaves some lesser amount of money to spend on other things. For example, if someone has a small budget to begin with, he or she might have to spend less on expensive forms of entertainment, or forgo it altogether, to make it to the next paycheck, thus trading off entertainment against subsistence. But if that person has a lot of money to spend, then it's strippers and cocaine all around (provided that you work on Wall Street). Although this means that certain trade-offs that might be considered fundamental can be alleviated or eliminated by accruing a large enough resource pool, animals that find themselves on the lower end of the energetic acquisition curve with little excess energy to spare are forced to make hard investment decisions that may come with serious fitness consequences.

The specific area of evolutionary biology that deals with the allocation of acquired energetic resources to specific fitness-enhancing processes and behaviors is called *life history evolution*. Life history is defined (somewhat tediously) as the study of scheduling of important events in an animal's lifetime. This means that it strives to understand why traits either are expressed or change their level of expression at particular times and for certain durations within an individual over its lifetime. But life history is much more than some dreary energetic bookkeeping with debits applied to certain traits at the expense of credits to others, because these resource allocation decisions can directly affect lifetime reproductive success.

Despite the repeated references to energy above, the currency of life history trade-offs—in other words, the metric that we use to measure them—is *residual reproductive value*. This refers to an individual's fitness in terms of its future reproductive potential. In other words, we judge how important a particular trade-off might be by considering how that trade-off affects an individual's ability to produce offspring, both now and at a later point in that individual's life. Since whole-organism performance traits not only are subject to such trade-offs by virtue of their (sometimes high) energetic costs but are also important determinants of individual reproductive success, we can predict that performance traits will be implicated directly in some key life history trade-offs. If we then go out and look for trade-offs involving performance, this is exactly what we find.

Please, Sir, I Want Some More

Of the various resource-hungry processes that animals must invest in, running an immune system is particularly pricey. The immune system beavers away in the background throughout an individual's entire life, cranking out cells and antibodies 24/7 and ramping up that activity significantly when mounting an immune response against an active infection. Because the immune system is so complicated, it's hard to say for sure how expensive it is, but estimates place it somewhere in the region of "extremely." For example, the energetic requirements of some mammals drop by around 30 percent when raised in sterile environments, and birds raised in similar environments (or supplemented with antibiotics) grow much faster than those raised in normal environments, where infectious agents are present.

Immune systems are a constant and unavoidable energetic expense so long as the individual wants to stay alive, which means that sick individuals are forced to divert resources from elsewhere to pay for increased immune activity during periods of infection. This causes trade-offs in immune expression with other energy-hungry life history traits, including performance. Although *Sceloporus occidentalis* lizards infected with the malaria-causing *Plasmodium* parasite show reduced stamina compared to uninfected individuals, this finding is likely accounted for by an accompanying 25 percent deficit in hemoglobin, and thus oxygen-supply capacity, in infected lizards. But performance declines in other animals exposed to pathogens can be more confidently ascribed to life history trade-offs.

By the early 1990s, scientists noticed that amphibians the world over are increasingly prone to mass die-offs and even extinctions caused by the disease chytridiomycosis. The pathogen, a fungus called *Batrachochytrium dendrobatidis* but more commonly known as chytrid, is globally distributed, but in some amphibian species it does not kill and instead manifests as a chronic infection, the consequences of which are poorly understood. One cost of such infections could be reduced performance ability over time as energetic resources are continually rerouted to pay the costs of elevated immune function that keeps the infection in check. Indeed, leopard frogs experimentally treated with the chytrid fungus show reduced jumping ability after eight weeks of lingering fungal infection. However, it need not take so long for immune function to impinge on performance abilities, and experimental activation of the immune system without a pathogen reduces maximum sprint speed by 13 percent within four hours in the lizard *Psammodromus algirus*, pointing to a clear trade-off between immune function and performance.

The point made earlier that performance is subject to life history trade-offs if it is expensive as opposed to merely costly bears repeating. It should also be apparent after a bit of thought that the distinction between a cost and an expense depends, at least in part, on the size of the available resource pool. Something that is usually expensive to invest in can thus become a lesser cost under resource-rich conditions. For instance, in yet another lizard (*Zootoca vivipara*), immune challenge decreased endurance ability only in individuals that had relatively low energy stores to begin with, whereas in lizards with larger energy stores, that trade-off was not apparent.

Because life history trade-offs can be masked in this way when food resources abound, a useful technique is to experimentally restrict the amount of food that animals have available to them. Doing so forces individuals to prioritize investment of their now-limited energetic resources, thereby exposing those priorities to scrutiny by weird people in lab coats. Although few studies have thus far taken this approach with regard to performance specifically, there is evidence that performance traits are indeed vulnerable to such induced trade-offs and that investment in performance traits is reduced when food is scarce. For example, juvenile green anole lizards raised on a restricted diet in my lab not only grew more slowly than lizards raised on an unrestricted diet but had much lower bite forces as adults than those lizards that ate as much food as they liked, even after accounting for the size differences between the diet treatments. This is because restricted-diet individuals could not afford to invest in the large adductor muscles that create biting ability, muscle being particularly expensive to grow and maintain.

There is also reason to believe that performance that is diminished through dietary restriction can be compensated for if environmental circumstances change and food resources again become abundant. Nick Royle of the University of Exeter found that green swordtail fish that experience an early but transient period of food scarcity as juveniles do not show reduced swimming performance as adults compared to swordtails that received abundant food throughout their lives (although they do experience other costs that are not compensated for). Thus, performance losses during one phase of life could be recovered at a later stage when food availability increases, and this is consistent with the notion that life history trade-offs are dynamic and dependent on the size of the current energetic resource pool.

Although it is easy to think of trade-offs in terms of one or two traits, such trade-offs can be complex. Altering resource availability prompts a variety of trade-offs in the organism, some of which involve performance directly and others only indirectly. What is more, these trade-offs can be different in males and females because of the potential for sex-specific selection on performance, as outlined in Chapters 3 and 4. A study involving dung beetles hints at the complexity involved. Female dung beetles lay their eggs inside brood balls that the mother makes from the dung of large animals. When those eggs hatch, the larvae feed on the dung of the brood ball itself, and so the mother determines the amount of food resources available to the larvae. Dung beetles are holometabolous insects, which means that each individual undergoes metamorphosis, changing from a larva to a pupa and finally to the adult beetle. Following eclosion (which is the term for the adult beetle emerging from its pupa), the overall shape and size of the beetle is fixed for life; beyond this point the dung beetle will never again grow larger and shed its exoskeleton, as some other arthropods do.

The horns of male dung beetles that arc shaped by sexual selection to be weapons or signals for use in male combat are outgrowths of the exoskeleton. This means that the size of both the beetle and its horns is determined primarily by the resources accumulated during the larval feeding stage (in addition to relevant genetic factors, about which more in Chapter 9). But although the exoskeleton never changes, the amount of muscle within can, and following eclosion the adult beetle undergoes a period of maturation feeding during which it grows large amounts of its soft tissue, including its musculature. We therefore have a situation where the size of a beetle's horn is set at the larval stage, but the amount of muscle, and thus presumably its strength, is a function of the adult environment. Given that I regaled you in Chapter 3 with the tale of a dung beetle who signals its strength to other males via the size of his horn, this potentially malleable link between horn size and strength now demands some explanation from the field of life history.

Leeann Reaney and Rob Knell at Queen Mary University of London undertook to discover this explanation in *Euoniticellus intermedius*, not at all coincidentally the very same dung beetle I discussed earlier. In *E. intermedius* each brood ball contains only a single egg, and so it is a simple matter to manipulate the amount of food available to the larva by changing the size of the brood ball. Reaney and Knell removed dung from some brood balls while leaving others untouched and also restricted the amount of food available during the maturation period for the beetles that emerged from the manipulated brood balls. They then measured a raft of traits in the adult beetles, including weight, strength, development time, the length of the elytron, or wing-cover, and (in males) horn size, the idea again being that by restricting resource availability throughout development, the trade-offs among all of these important traits would be revealed.

The results showed that the influence of that resource availability was straightforward for females: larger brood balls begat larger beetles, and those larger beetles were also stronger (fig. 8.1a). But the results for males showed a complex network of links and trade-offs among the measured traits, such that the development of horns in the males diverted energetic resources away from the other measured traits resulting in a net *negative* relation between brood ball size and strength (fig. 8.1b)! The positive link between horn size and strength remained, however, as male beetles that were able to grow bigger horns also invested energy in growing tissues associated with strength. Studies such as this illustrate the complicated and sometimes

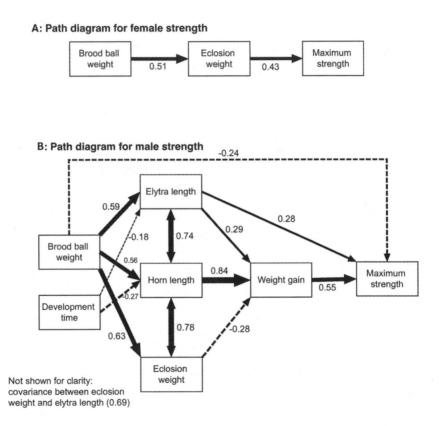

Fig. 8.1. Path diagram summarizing the relations among factors affecting strength in (a) female and (b) male *E. intermedius* beetles. The resource allocation trade-offs involving the male horn mean that brood ball size has a weakly negative correlation with male strength. From Reaney and Knell 2015.

counterintuitive nature of certain life history trade-offs, as well as the challenges facing those of us who hope to understand them.

No Pain, No Gain

Although dietary manipulations are a valuable means of exposing life history trade-offs, manipulating animals to invest in performance (just as we force them to invest in immunity via immune challenge or activation) would be especially useful. But how to incentivize animals to allocate resources to performance specifically? Jerry Husak of the University of St. Thomas realized that he might be able to do so by training green anoles in particular performance tasks.

The rationale behind this approach is that exercise prompts several changes in the animal that are specific to the type of training undertaken. For example, resistance training in humans increases muscle mass, whereas endurance training increases oxygen supply via changes such as increasing heart size and hematocrit. The *exercise response* therefore involves investment in the factors that influence whole-organism performance. If that investment is large enough, it could also drive trade-offs with other traits.

The notion that other animals besides humans have an exercise response may seem odd, but there is evidence that amphibians, birds, and especially fish can increase their maximum performance abilities through dedicated use. In guppies, exercise may even be ecologically relevant, as female guppies that are forced to escape constant harassment from males become more efficient swimmers than females that are exposed to lower levels of harassment. With regard to lizards specifically the evidence for a training effect has historically been ambiguous, but Husak suspected that the training regimens that had been applied to lizards previously were not appropriate to elicit an exercise response. Armed with a pet treadmill and an army of undergraduate minions, Husak's lab was soon training green anoles for endurance.

Husak quickly found that by applying the right training regime green anoles can be trained for endurance, just like humans can, with treadmill-trained green anoles showing significant improvements in loco-motor endurance abilities after eight weeks of workouts. But as interesting as this saurian training response is, the other changes accompanying it are even more so. Female endurance-trained anoles lay fewer eggs than un-trained females, suggesting that the training response diverts resources away from reproduction. Females lay even fewer eggs if endurance training is combined with a dietary restriction regime. Investment in the exercise re-sponse also suppressed immune function in trained anoles, a finding that is particularly intriguing in light of similar findings that strenuous exercise and overtraining of human athletes renders them susceptible to infection.

Husak's work shows that exercise training can indeed force life history trade-offs among fitness-related traits, and so even though we may not think of lizards as being dedicated runners in the same way that those training for marathons are, the expression of their performance abilities is not free and comes at a very real cost.

It's Not the Years, It's the Mileage

Resource allocation doesn't just happen at one or even a couple of stages in an animal's lifetime. Rather, it is a dynamic and ongoing process that depends not only on the current environmental conditions and selective context but also on the type and duration of the allocation trade-offs that have gone before.

Selection can act differently on animals of varied ages, such that the same trait that is beneficial when an animal is young can become selectively neutral or even detrimental later in that individual's life. In collared lizards, for example, sprint speed predicts survival in juveniles but does not influ-ence adult survival (although it does predict reproductive success in adult males but not females). It would thus benefit animals to be able to mitigate, abolish, or even reverse investment in certain life history trade-offs at differ-ent life stages. Viewed from the perspective of an animal's entire lifetime, these allocation trade-offs can affect not only how long an animal lives but how the expression of a given trait changes over the animal's lifetime — in other words, aging.

To most of us, aging is synonymous with *senescence*, defined as the age-related decline in the expression of fitness-related traits due to the progressive loss of physiological function. This is how we experience aging past a certain point in our lives, and it is clearly evident in age-related patterns of athletic performance in humans. Among the various athletic events in which certain proportions of the human population compete for some reason, every single one exhibits a significant decline in level of performance past a certain competitor age. The bad news for those of us in our thirties and beyond is that we are already on the downward slope of senescence in terms of almost all sports that have been considered so far. There is no good news.

But not all performance traits decline in an identical fashion, nor does everyone experience such declines equally. Weightlifting exhibits the fastest decline with age, and jumping and weightlifting decline most of all in women relative to men. The proximate reasons for senescence amount to the breakdown as an organism ages of the various cellular processes that are essential for maintaining proper cellular function. Those mechanisms are interesting and important, and at any given time newspapers around the world are running approximately forty thousand articles with titles that are variations on "Who Wants to Live Forever?" reporting on the recent attempts to arrest this deterioration and "cure" aging. But even as aging was becoming a biomedical question, evolutionary biologists such as George Williams and the influential thinker Peter Medawar were interested in why animals have not evolved to maintain those cellular mechanisms and ensure that senescence does not occur. From this perspective, the key evolutionary question "Why do we age?" could more specifically be phrased as "Why do we senesce?"

The reason these questions are different is because senescence and aging are not necessarily synonymous. The fact that we humans are inexorably prone to senescence with all of its depressing existential implications has blinded many of us to the reality that senescence isn't the only possible aging pattern in nature. Some animals practice a life history strategy called *terminal investment*, whereby they allocate more and more resources to a particular trait as their survival prospects decline and their inevitable death looms ever larger before them. This means that those traits do not senesce with age but increase their expression as the animal grows older.

This strategy has its roots in what is called the *disposable soma* theory of aging, and it implies that patterns of aging can be altered by life history trade-offs. The logic of this strategy is that because death is coming—very soon in fact, perhaps due to the mounting probability of being eaten by something larger than you the longer you are alive—it makes little sense to continue investing in somatic maintenance (that is, making sure that all of those cellular repair mechanisms are running smoothly), which becomes increasingly futile with each passing day, when a much better use of your energy would be to invest it into things that can increase your fitness, such as attracting mates and having as much sex as possible, in the little time you have remaining. Terminal investment thus alters the aging trajectories of certain traits such that their expression increases with age even while longevity itself is truncated. Both senescence and terminal investment can apply to and indirectly affect (via trade-offs) athletic performance.

Fish under Troubled Waters

Guppies are best known as popular aquarium fish, but they are also a popular study organism in evolutionary biology. As well as being model organisms for the study of sexual selection (see Chapter 3), guppies have been the subject of one of the most important life history evolution experiments ever conducted. Fortunately for us, this long-term study involves athletic performance explicitly.

David Reznick of the University of California, Riverside, and collaborators have been studying the life histories of stream-living guppies in Trinidad since the 1980s. This research has focused on guppy reproductive scheduling, as well as their longevity, and has revealed key insights into how life histories can be shaped by different ecological milieus. The researchers manipulated predation rates in natural populations by excluding predators (the pike cichlid, specifically) from some streams but not from others. By altering the risk of predation alone, Reznick and his team evoked significant and strikingly fast changes in guppy morphology, coloration, age of

maturity, longevity, and performance between the two treatments over relatively few generations. Within only two years, guppies in the high-predation
sites died sooner, matured earlier and at a smaller size, devoted more energy toward reproduction, and showed higher escape speeds compared to
guppies from the low-predation sites. In addition to being faster, the high-
predation site guppies also evolved to be less colorful, probably because
being colorful and attractive (and therefore conspicuous) increased their
risk of predation.

Faced with a shortened life expectancy driven by the high probability
of finding themselves in the belly of a pike cichlid, those guppies in the
high-predation sites shifted their resource allocation toward reproducing as
much and as soon as possible, and also to escaping so that they could survive
long enough to do so. Importantly, subsequent experiments rearing guppies from high- and low-predation populations in the laboratory showed that
guppies from the high-predation sites didn't live as long as those from the
low-predation populations *even when there were no predators present at all,*
suggesting that this strategy had a fixed, genetic basis. By evolving to divert
energy investment away from cellular maintenance and into reproduction
and escape, these fish had curtailed their own longevity (which is probably
fine because there's no point in planning for old age if you can be confident
you will never reach it), as predicted by the disposable soma theory.

And there's more: although guppies from high-predation populations
were faster than guppies from low-predation populations when they were
young, they also senesced in swimming performance earlier and more rapidly than low-predation guppies! Thus, while it appears that investing in
performance when young can be a viable strategy, the costs of deferred senescence cannot be delayed forever.

Live Long or Prosper

Guppies are just one prominent example of a ubiquitous trade-off between
reproduction and survival, the importance of which is arguably not always
fully appreciated even today. It's probably a safe bet that everyone has heard
the phrase "survival of the fittest" at some point. While undeniably pithy,

this maxim directly equates survival with fitness in the evolutionary sense: the fittest animals are, it implies, those that survive the longest. The particular way this phrase is worded also directly implies the reciprocal, which is that living (surviving) longer translates directly into fitness.

This isn't necessarily true. I've already said several times that fitness is about reproduction, not survival, and the fittest individuals are therefore those that produce the largest number of offspring in their lifetime *regardless of how long that lifetime actually is*. This means that an individual that invests so heavily in reproduction that its own longevity becomes compromised as a result, and that consequently dies young but with bounteous issue, is fitter than an individual that invests less in reproduction, lives much longer, but produces fewer offspring. Males of certain animal species do exactly this, choosing sex and reproduction over survival.

Male crickets gain reproductive success by making themselves attractive to females, and they will invest a great deal of energy to do so. Attractiveness in *T. commodus* is all about calling, and males who call more often attract many more females than males who call much less. But because calling is so energetically expensive, investing heavily in calling prompts life history trade-offs with other key traits. If given the appropriate resources, male crickets call so much that they reduce their lifespan significantly, but because they boost their attractiveness by doing so, they can nonetheless realize high fitness. Males who adopt this "live fast, die young, leave a whole bunch of kids" strategy do so by investing terminally in calling, such that their calling effort does not senesce but increases as their survival prospects decline. Males who do not adopt this strategy enjoy much longer lifespans, which they presumably spend at home playing video games with their other single, offspring-poor male cricket friends.

I've talked about *T. commodus* before, and readers who have yet to begin the slow but inevitable descent down the slope of senescence toward the Lethean fugue that awaits so many of us may remember this terrible creature[3]

3. I worked on this animal during my time as a postdoctoral fellow in Rob Brooks's "Sex Lab" at the University of New South Wales in Sydney. Almost immediately, I developed a deep and abiding hatred for all cricketkind thanks to the crickets' obnoxious standards of personal hygiene and overall bad attitude toward the experiments I

from Chapter 3, when I described the genetic relation between attractiveness and performance. Given that attractiveness and performance show a negative genetic correlation in male black field crickets, and taking into consideration the energetic expense of calling, one might reasonably predict a negative relation between calling effort and jumping ability, such that terminal investment in calling results in rapid senescence in jumping ability.

Predictably, the reality is not that simple. Jumping in male *T. commodus* shows little evidence of age-related change, possibly because it is buffered against life history trade-offs by the very same elastic storage mechanisms that allow them to jump in the first place. Bite force, however, most certainly does show evidence of aging in these crickets, although here context as well as the sex of the animal appears to be extremely important. Female crickets who remain virgins their whole lives show an aging pattern for bite force that is consistent with terminal investment (that is, bite force increases with age), as do male virgins. Males who are given the opportunity to mate once a week, in contrast, don't change their bite force expression as they get older, but females under the same mating regime senesce in bite force! So it appears that not only are patterns of performance aging flexible in *T. commodus* depending on what other expenses may be incurred at any given time (mating being especially costly for various reasons), but they can also differ depending on the type of performance in question.

Aging in Other Athletic Animals

Like many of the issues that I have touched on only briefly in this whirlwind tour of performance in the animal kingdom, aging is a complex and fascinating topic, the surface of which has borne but the briefest of scuffs from

wanted to do with them. Several colleagues have pointed out to me that it's not usual for biologists to talk trash about their study organisms at every opportunity; in my defense, most animals are not nearly as nasty as crickets. They are, however, superb organisms for studying life history, and researchers in the Sex Lab and elsewhere have done excellent work on how these crickets invest their resources and on how they age. In fact, crickets are so useful for asking these questions that I still work on them ten years later, despite my dislike for them.

our inspection here. Work on aging in nonhuman animals focuses heavily on traits that are tied to reproduction, and understandably so. But the result is that few researchers have thus far turned their attention to aging in performance, and so we still have much to learn about how and why animals alter their energetic investment in performance at different points in their lives, and what the consequences of those decisions are for performance aging.

Many of the challenges are logistic. An ideal organism for studying aging is something with a relatively short lifespan that can be raised or otherwise maintained in the laboratory for its whole life and is small enough to be kept in large numbers. Few noninsects meet these criteria, and even many insects are less-than-ideal lab animals. Studying aging in animals with long lifespans is difficult and often involves drawing inferences from cohorts of animals of differing ages, as opposed to following individuals throughout their lives, as can be done with guppies or crickets.

Because much of what we do know about the aging of performance therefore comes from small animals that can be kept easily in the lab, it is reasonable to ask, as some have done, whether results gained in a laboratory environment removed from many of the factors that drive life history trade-offs are relevant to animals living in the wild. For example, grey mouse lemurs in captivity show weaker grip strength compared to wild animals, and they also display earlier senescence in their grip strength compared to wild lemurs, possibly due to increased mortality in wild animals. Yet other studies have turned the premise of this chapter on its head, asking, not how aging affects performance, but how performance activity over an animal's lifetime affects its aging trajectory and overall lifespan. This has long been an area of interest in humans, where exercise has been linked to numerous health benefits, and aerobic capacity is considered to be the most important predictor of lifespan after (not) smoking.

Findings from the animal world are more variable. Suppressing flight behavior extends longevity in some small flying insects, whereas forcing frequent flight hastens senescence of flight performance in fruit flies. By contrast, endurance activity in mice that are genetically modified to senesce faster than normal mice reverses many of the systemic aging effects otherwise seen in sedentary individuals, and even delays mortality; yet, when regular mice are bred for high endurance ability, the onset of senescence is delayed in females, but occurs earlier in males. It is clear that we have a great deal to learn about performance aging, and the mice studies offer just a hint of the challenges and promise offered by genetic approaches in particular.

NINE

Nature and Nurture

Throughout this book I have tried to explain how particular performance traits work and why they have evolved. But although natural and sexual selection are important drivers of evolutionary change, they are only part of the story. For selection to cause evolutionary change in a trait, selectively advantageous traits have to be passed on to the next generation through some mechanism. In other words, traits that are acted on by selection (that is, phenotypes) must be at least partially controlled by heritable elements such that individuals that inherit those particular elements express the phenotype in a similar way to their parents.

Furthermore, natural selection operates by choosing from among several possibilities the particular phenotypes that work best within a particular environment. Because selection cannot choose the best phenotype if all of them are identical, there has to be *variation* in expressed phenotypes within a population. It therefore follows that there must also be variation in those heritable elements that affect phenotypic expression—if there isn't, then all phenotypes will be the same across generations, and there can be no variation for selection to act on.

Ever since the work of Austrian monk, biologist, and pea enthusiast Gregor Mendel in the late nineteenth century, we have known that these elements that both affect phenotypic variation and allow phenotypes to be inherited are genes—specific sequences of nucleic acids that make up the DNA of all living organisms. Because phenotypic (and therefore genetic) variation is so important to evolution, we need to have some idea of the extent of such variation in performance traits if we hope to understand the processes driving performance evolution. A variety of methods have been deployed to assess the genetic basis of performance traits ranging from experimental selection and quantitative genetics to genomewide association analyses. Despite these approaches, our understanding remains woefully inadequate, and much of this chapter will therefore dance nimbly around what we don't know. Although the concepts underlying what I am about to tell you are not difficult to grasp, genetics in general is not for the faint-hearted, and you might want to take a break from time to time throughout this chapter for a stiff drink or twelve-hour nap.

Nature ~~versus~~ via Nurture

At this point, I think I should make a confession. Even though I find performance and exercise endlessly fascinating, I don't particularly enjoy partaking in it and do so as seldom as I can get away with. This is because getting into and staying in shape requires effort and cuts into time that I could otherwise spend not doing that. The most sports-related activity I get therefore comes from lying on the couch and hurling criticism at world-class athletes who are my physical and mental superiors. I nonetheless drag my creaking and aging body to the gym several times a week and force myself to run and move heavy things around for one primary reason: I have a family history of type II diabetes and blood pressure issues.

Both type II diabetes and blood pressure problems arise from *genotype-by-environment interactions* (GxE). This is the technical genetics term for the situation where having a particular suite of genes or genetic variants predisposes one to expressing a certain trait, whether it is diabetes or something else, but (and this is the important part) *the likelihood of or degree to which that trait is expressed is strongly dependent on the environment.* (Here, "environment" refers to anything that isn't genetic.) This means that if I sit around in front of a computer all day not exercising and subsisting entirely on beer, potato chips, and ramen noodles (a state better known to scientists as "being in graduate school"), I can very likely look forward to a not-too-distant future filled with insulin shots and sugar-free everything. But

I can partially mitigate my chances of that happening by eating properly and forcing myself to partake in athletic activities that I don't enjoy. (I'll be really annoyed if I get diabetes anyway.)

That the expression of a gene or group of genes is dependent on the environment in which it is expressed is an absolutely crucial issue appears to have been lost on those newspapers, magazines, websites, and news channels that are enamored of headlines and soundbites such as "Nature versus Nurture: Is [Obesity/Genius/Aging/ADHD/Fussy Eating/Sexual Deviancy] All in Your Genes?" I will save you the trouble of reading any of those and reveal that the answer is always, "No, it isn't." Nature versus nurture is a false dichotomy. *Both* an individual's genetic makeup *and* an individual's environment almost always influence how those genes are expressed, which means that many phenotypes—most, in fact—are susceptible to GxE interactions that can have important implications for their evolution. Performance traits are no exception, but to understand why, we first need to understand something about how inheritance works.

Connecting Genotypes to Phenotypes

If we look at the range of phenotypes that an organism expresses, we see that some of them are easily categorized. Human eyes, for example, are limited to a handful of discrete colors, such as green, blue, brown, or gray. Similarly, birds lay a discrete number of eggs—one, three, or four, but never 1.5 or 2.8. These *categorical traits*—so called because we can place their expression into distinct categories—are described by simple genetic rules of inheritance that Mendel derived from experiments he performed on pea plants back in the day.

Briefly, by breeding and cross-breeding pea plant varieties that varied in the expression of various discrete traits (producing peas that were either round or wrinkled, yellow or green, and so on), Mendel was able to measure the ratios in which these traits were expressed in subsequent generations, such as three yellow pea plants to every green plant. The constancy in these ratios across generations led him to realize that these discrete traits were under the control of specific heritable elements that we now call genes, and that the phenotypes of offspring of known parentage could be predicted based on the parental phenotypes.

This happens because offspring get one set of their genes from the mother and another set from the father. The specific combination of parental alleles (different versions of the same gene) determines how the phenotype is expressed. Let's take pea color as an example, and let's say that there are two alleles controlling pea color: Y, which makes peas yellow, and y, which makes them green. A given plant has two sets of chromosomes (one from its mom and one from its dad) and thus a specific combination of color alleles. Specifically, it may be YY, Yy, yY, or yy for pea color. Because of a property of alleles called *dominance*, YY, Yy, and yY individuals all produce yellow peas, whereas only yy individuals produce green ones. We therefore say that Y is *dominant* over y and that y is *recessive*. This property of dominant and recessive alleles gives rise to the three-to-one ratio in yellow versus green pea color and similar ratios in other discrete traits that Mendel observed in successive generations of offspring within a family.

This brief sketch does not do justice to even simple Mendelian inheritance, but the key point is that the discrete phenotype (pea color) can be linked directly to a specific gene or gene combination (called a genotype). Many other phenotypes, however, are under the control of multiple genes or alleles. Eye color is actually determined by more than fifteen genes, which is why we can discern different shades of blue or brown eyes. But as the number of genes involved in determining a phenotype grows, those Mendelian ratios become less and less obvious, until eventually they are subsumed beneath a sea of phenotypic variation. At this point, those discrete categories can no longer be delimited; variation has become a continuous spectrum from one extreme to the other.

If we consider height as an example, humans are not just short or tall but of varying heights. As a consequence, the inheritance of continuous phenotypes such as height cannot be understood using traditional genetic rules such as those that Mendel devised for his pea plants. For the same reason, the identities of the specific genes affecting height, of which there could be dozens, hundreds, or many thousands, and a large percentage of which might interact with one another to form intricate gene networks, cannot be discerned through simple breeding experiments. For continuous traits the problem is no longer discriminating among the genes of different colored peas and more akin to separating out the ingredients of pea soup.

The large numbers of genes contributing to variation in continuous phenotypes means that determining their mode of inheritance is a challenge. But it gets worse. As ranted above, gene expression can be altered by the environment such that different environmental conditions can result in different phenotypes from the very same gene or set of genes. A simple example gives us a hint as to the potential mechanisms.

In cats, a *point mutation* (that is, a change in just one DNA nucleotide "letter" making up a DNA sequence) in the gene that produces an enzyme called tyrosinase causes that enzyme to be built containing a slightly different set of proteins. The resulting tyrosinase is faulty. Normal tyrosinase catalyzes the production of the pigment melanin that is responsible for dark coloration in most of the animal world. Completely broken tyrosine sabotages all melanin production and leads to albinism, and thus blue-eyed white cats (which also tend to be deaf and to have poor eyesight because melanin also performs other functions in the brain and during development). The faulty tyrosine caused by this point mutation, however, is sensitive to temperature—it works fine in the cold but not at all when it is warm. This mutation gives rise to Siamese cats, whose cooler parts of the body—the snout, ears, limbs, tail, and especially testicles—are colored darkly, whereas the warmer parts of the cat produce no melanin at all and lack pigment entirely.

Every Siamese cat is therefore a frustrated black cat because of the interaction between its faulty tyrosinase gene and its (in this case thermal) environment. Indeed, raising a mutant kitten in a cold room induces it to grow dark hair, whereas a warm mutant kitty becomes white-haired and blue-eyed.

You Are What You Eat

Most phenotypes that we see in the animal world are therefore predicated on networks of interacting genes, any one (or all) of which may also be subject to GxE interactions. Thus, two individuals can harbor the same genes but express different versions of a phenotype depending on the environment in which they find themselves. This phenomenon of trait expression being affected by changes in one or more environmental factors is called *phenotypic plasticity*. The thermal dependence of ectotherm performance discussed way back in Chapter 5 and the exercise response from Chapter 8 both constitute cases of performance plasticity, where performance changes in response to environmental influence (that is, temperature and activity level, respectively).

Sources of phenotypic plasticity are myriad. Although the obvious environmental factors such as temperature and moisture can affect phenotypic expression, so too can others, such as diet or the hormonal milieu inside the mother's womb or eggs that she lays, and even in some cases the short-term effects that her male mate has on her own reproductive tract! Of these sources of performance plasticity specifically, dietary influences may be especially important, not only via variation in resource availability, as mentioned in the previous chapter, but also through diet quality.

The field of nutritional geometry concerns itself with investigating how ratios of carbohydrates, fats, and protein in animal diets affect the optimal expression of various traits, and a related approach called *ecological stoichiometry* similarly tests how trait expression is influenced by the dietary availability of micronutrients or elements such as nitrogen and phosphorus in different animal species. We are all familiar with the strategy of carb-loading, whereby endurance athletes might eat high-carbohydrate foods before competition to ensure a sufficient store of glycogen to power sustained activity, the idea being that this strategy can enhance endurance capacity. Animals have their own dietary preferences tailored to their various ecological and selective needs and prefer to eat foods that meet those intake targets in terms of diet quality (such as a preferred ratio of fats to carbohydrates to protein) rather than mere quantity. This means that the specific diet that animals evolve to use could constrain or enhance endurance performance, for instance, depending on which traits are being optimized by that diet. An animal that relies extensively on endurance in its day-to-day existence might find itself in a bind if suddenly forced to subsist on a high-protein, low-carbohydrate diet.

Because these dietary requirements are so important, some animals have evolved the amazing capacity to regenerate stores of their necessary fuel type almost independent of the specific dietary resources available in the environment. The desert western chestnut mouse lives on arid Barrow Island off the western coast of Australia and subsists on a nutrient-poor diet of spinifex grasses. This animal is very small, and its stores of glycogen, which power burst activities such as rapid predator escape, are correspondingly sparse, which means that the little chestnut mouse can and does exhaust all of those stores in a few consecutive spurts of rapid sprints to exhaustion. Laboratory experiments by researchers at the University of Western Australia reveal that this mouse can restore its glycogen stores that power sprinting to preexercise levels within fifty minutes of exercising to exhaustion—without eating any food at all! It is very unlikely that western chestnut mice are biological perpetual-motion machines that will soon be powering our cities through advanced mouse-wheel technology, and so these animals must be reconstituting their glycogen stores from other potential sources, such as fats, lactate, or (less efficiently) proteins, and doing so with startling rapidity.

Alaskan sled dogs are much larger animals but will happily run themselves into the ground forever, achieving daily distances of 250–300 km (155–186 mi.) at average speeds of 15–20 kph (9.3–12.5 mph) day after day under conditions that can euphemistically be described as difficult. Researchers at the University of Oklahoma found that trained Alaskan sled dogs that ran 160 km/day (just under 100 mi./day) for four consecutive days on a limited carbohydrate diet not only replenished their glycogen stores much faster than control dogs that did not run but did so faster over the course of the last three runs compared with the first one. This finding mirrors studies on humans showing that human long-distance runners can adjust over time to high-fat diets that are normally considered suboptimal for supporting endurance, albeit not as quickly as the sled dogs. Similarly, the structure and function of the digestive tract are also plastic in some animal species, and change in response to dietary shifts. This means that not only is performance itself plastic and affected by environmental factors but there is also plasticity in those underlying factors that affect performance plasticity!

Inherit the Wind

Sled dogs are, of course, bred by humans for running stupidly long distances at absurdly fast speeds, just as we impose artificial selection on other animals such as racehorses and greyhounds to push them to their athletic limits. Such directed artificial selection mimics the natural selection pressures to which animals in the wild are subject, with the important distinction that natural selection has no long-term end-goal in mind and is concerned only with what increases fitness right here and now. Natural selection therefore fluctuates in direction and intensity, whereas human-imposed selection is directed constantly toward a specific endpoint.

But artificial or experimental selection is greatly facilitated if we can differentiate between the genetic and environmental components (or, if you prefer, the nature and nurture components) of the trait of interest. This is because *only the genetic components* can be passed on from parent to offspring to offspring to offspring and so on. Parent and offspring resemble each other more closely than either does an unrelated individual because they have more shared genes in common with each other than with non-relatives. Now, if an offspring is born into the exact same environment as its parents, then the resemblance between parent and offspring is inflated because they now share *both* a genetic *and* an environmental aspect to their phenotype.

As evolutionary biologists, we are particularly interested in the parent-offspring resemblance that remains even if the offspring finds itself in an entirely different environment, which can happen frequently over evolutionary timescales, because it is this genetic component of the phenotype that is subject to modification via natural and sexual selection. We call this genetic component the *additive genetic variation*, and we usually express the influence of additive genetic variation on a given phenotype as a convenient metric called *heritability* (h^2).

Heritability has baffled students for generations, so I'll take its meaning and interpretation slowly here. It is a dimensionless number ranging from zero to 1 that tells us how much (specifically, what proportion) of the variation in our phenotype of interest (which might be affected by dozens or hundreds of genes) is accounted for by additive genetic variation alone.[1] To be clear, we cannot look at something like a six-foot-tall person and say, "Genes are responsible for four feet of height, and environment for the remaining two," because the relative importance of the additive genetic and environmental influences become apparent only at the population level (which is the level at which we can observe and measure how individuals differ from one another). *Heritabilities are thus characteristics of populations*, not of individuals, and if we measure h^2 for, say, sprinting, this tells us about the extent to which individual differences in sprinting are under genetic control among members of the focal population.

An h^2 of 1 therefore means that all of the differences in sprinting ability among individuals are of the additive genetic kind—because members of the population each have different alleles or groups of genes that affect sprinting—and there is thus little to no plasticity. An h^2 of zero, on the other hand, means that there is likely little to no genetic variation for sprinting ability and that everyone in the population has roughly the same complement of genes affecting sprinting. It does *not*, however, mean that sprinting has no genetic basis and cannot be passed onto offspring! A trait with a heritability of zero is nonetheless *in*heritable, which is, I suspect, the source of much confusion; but even though such a trait can be passed on to offspring, it cannot evolve via natural selection, as we shall see. A zero heritability also suggests that all of the differences in trait expression that we see among individuals within that population are caused by plastic environmental effects.

Heritabilities for performance traits differ greatly depending on the animal and the type of performance in question. *Zootoca vivipara* lizards, for example, have an endurance heritability of 0.4, which means that about 40 percent of the variation in endurance in *Z. vivipara* is additive genetic in nature, and the remaining 60 percent is accounted for by environmental effects (things like diet, condition, activity, and so on). Heritability for sprint speed in this same species, however, is very close to zero. Heritabilities for

1. This is the definition of *narrow-sense heritability* specifically. We can also talk about *broad-sense heritability* (H^2), which includes other types of nonheritable genetic effects in addition to additive genetic ones and is less useful.

other performance abilities in nonlizard species tend to be on the similarly low to moderate side: 0.3 for jump power in *Teleogryllus commodus* crickets, 0.21 for flight duration in the fly *Drosophila aldrichi*, 0.15 for mean take-off acceleration in the butterfly *Pararge aegeria*, and so on.

Heritabilities alone don't tell us very much, but we can make some tentative inferences based on their relative magnitude. Traits that are under strong selection pressures often exhibit low heritabilities because very strong selection tends to erode additive genetic variation—in other words, if selection is working especially hard to weed out inferior genetic variants, then it usually results in populations where everyone has very similar genetic profiles and thus low heritabilities for the trait under selection. We might infer that traits with especially low heritabilities (such as *Z. vivipara* sprinting) are under particularly strong selection, and thus of special importance to that animal's fitness, as expected based on our observations that such traits are important for escaping predators or acquiring mates, or both. We have to be careful, though, because a low heritability could also mean that the trait is under a relatively strong environmental influence (remember, h^2 is the ratio of additive to total phenotypic variation, including environmental effects, which means that low h^2 could be the result of *either* very low additive genetic variation *or* especially large environmental variation).

Heritabilities let us make inferences about how selection might have operated on that trait in the past, but they also allow us to do something even more useful, and that is to predict evolutionary change in the future through an equation called the *breeder's equation*. If you have a crippling fear of mathematics, then at this point you might be feeling some apprehension, but I assure you, this equation could not be simpler. It states that, for a given trait, the response to selection (R) is equal to the product of the strength of selection on that trait (s) and its heritability (h^2):

$$R = h^2 s$$

In other words, if we know exactly how selection is currently acting on something like jump performance (that is, whether and to what extent jumping ability is favored or disfavored by selection) and we know the heritability of performance, then we can figure out how jump performance will change in response to that selection (that is, whether and to what extent it will increase or decrease in the next generation). The selection differential (s) therefore represents numerically the strength of any of those natural and sexual selection pressures that I have talked about throughout this book that might drive evolutionary change.[2]

This equation states explicitly that there will be no response to selection (R)—in other words, no phenotypic change in the subsequent generation—if there is no selection operating ($s = 0$) or if there is no additive genetic variation for that trait ($h^2 = 0$). Again, the key point to remember is that h^2 is explicitly about variation, and it is worth emphasizing that traits with no genetic variation cannot evolve because they are passed on to offspring unaltered by selection, regardless of how strong that selection may be.

Evolving Hyperactive Mice

In my evolution class at the University of New Orleans, I introduce the breeder's equation as "the most important equation no one has ever heard of" because it forms the basis of our understanding not only of evolutionary change in nature but also of all forms of artificial selection. Humans have altered the strength and direction of selection (s) on organisms as diverse as wheat, cabbage, cattle, and sled dogs (to name only a few) for thousands of years to mold their morphology, life history, and behavior to our needs and preferences. The breeder's equation explains the success of those ancient artificial breeding programs, and its predictive ability (in combination with another important genetic metric called the *breeding value*, which represents the purely genetic aspects of trait variation inheritable from each particular individual—nature minus nurture entirely) allows for unprecedented refinements in modern breeding programs. I once had a fascinating conversation about animal breeding in a bar with a man who turned out to

2. The technical meaning of s is the difference between the mean trait value of the initial population preselection and that of the individuals from that population who are selected to be the parents of the next generation. Viewed in this way, heritability is the fraction of the change in phenotype that is inherited by the offspring.

be a retired farmer from Iowa. It was not surprising that he knew more about breeding values than I did.

Breeding programs also have an important place in modern evolutionary biology. Applying artificial selection techniques to traits of interest and observing how they respond to that selection over multiple generations is a powerful approach, albeit one that cannot be applied easily to organisms with long generation times. To do this, researchers choose individuals in each generation that are within some predetermined range for the trait of interest—those stalk-eyed flies in the seventy-fifth percentile for eyestalk length, for example—and use them as parents of the subsequent generation, thereby manipulating the s component of the breeder's equation. By selecting repeatedly over numerous consecutive generations, biologists can effect evolutionary change in that lineage and compare it to changes in control lineages that do not undergo selection, as well as to other lineages that may be selected in different ways. This is a useful technique because it can reveal links among traits that we might otherwise never have known existed—all you have to do is change the direction or strength of selection, and see what happens. Although the way I have phrased this sentence might suggest that this is an easy thing to achieve, it is not, and selection experiments are both time consuming and logistically challenging.

Since 1998, Ted Garland and his many collaborators have conducted a remarkable long-term selection experiment on voluntary mouse-wheel running with the aim of determining "the genetic and physiological bases of voluntary wheel-running behavior and simultaneously to study the correlated evolution of behavior and physiology." This experiment has been running for more than seventy generations so far and has yielded a treasure trove of information related to how and why selection for locomotor ability evokes changes in the overall way that mice are built and go about their lives.

Garland and his colleagues developed an ingenious protocol whereby the mice measured themselves for locomotor capacity via a mouse wheel included in each individual enclosure. The investigators thus based their selection for high levels of wheel running on those mice that chose to spend more time on the mouse wheel than other individuals. Because experimental evolution experiments require replication to ensure that results are not due to chance or random events, Garland's group maintained four duplicate lines of high voluntary wheel-running mice as well as four control lines that did not undergo any selection.

The results of this experiment are fascinating. Not only do high wheel-running mice increase their wheel running ability, but they also exhibit a variety of other changes. Selected mice today are faster, smaller, leaner, grow more slowly, and run more cheaply than control mice. They also have larger hearts and different mitochondrial enzyme levels and muscle fiber types than controls. Although some of these changes are as expected, others are more surprising. For example, high-wheel runners exhibit overall larger brains and, intriguingly, larger midbrains. Indeed, it appears that in addition to changing the overall size of the brain, selection for high-wheel running has changed other aspects of the nervous system, such that the neurobiological profiles of the brains of high-running mice closely resemble those of humans with attention deficit hyperactivity disorder (ADHD).

Because the experiment selects specifically for voluntary wheel-running activity, what is primarily different about high-wheel running and control running mice is their motivation to run. Selected mice are more active than control individuals and move around their cages more often (although not necessarily faster). Subsequent experiments show that the brain's reward system, specifically the component that is regulated by the neurotransmitter dopamine, is different in high-wheel runners, and in the same way that the brains of humans with ADHD are different to those of people without ADHD. Treatment of the wheel-runners with the ADHD medication Ritalin, which affects those areas of the midbrain involved in dopamine release, dampens their hyperactivity, just as it does in humans, and activates parts of the brain differently in selected and control mice. But dopamine is a marvelously complex molecule that also has important effects on addiction, among other phenomena, via its role as a signal feedback for predicted rewards. In this case, mice are rewarded for their running efforts with a hit of dopamine, similar to how the human brain is flooded with dopamine during orgasm (but presumably to a lesser extent). In effect, the high wheel-running mice are addicted to exercise at a genetic level through selection on the brain's dopamine-based reward system, and selected mice that are prevented from running show patterns of brain activation similar to experimental junkie mice that are denied their daily fix of cocaine, nicotine, or morphine. (This mechanism is, however, probably not the same as that involved in the runner's high experienced by humans.)

This remarkable experiment shows not only that physiology and morphology can and do evolve under selection for locomotor ability but also that behavior and its neurobiological underpinnings evolve in concert. Indeed, links between brain size or capacity and performance abilities have become a recurring feature in the evolutionary literature, leading to suggestions that brain size in some groups of animals, such as mammals, has been driven at least in part by evolutionary changes in locomotion, in addition to other cognitive factors.

Nearly Null Genetic Subspace and Other Words

That Garland and colleagues were able to obtain an evolutionary response (R) to their experimental selection regime (s) depended on voluntary wheel running initially having a non-zero h^2. However, most of this response appears to have occurred early on in the study, and there has been little appreciable change in voluntary wheel-running duration in the forty or so generations succeeding generations 16–28. This very likely corresponds, at least in part, to the erosion of genetic variation under strong selection that I referred to previously, leading over time to a low h^2 and thus a greatly reduced response to selection.

Lack of additive genetic variation constitutes a genetic constraint on trait evolution that can be overcome only by the introduction of new genetic variation into the gene pool. This can come about through mutation or outbreeding with new individuals from elsewhere harboring a different complement of alleles influencing the trait of interest. Racehorses have long presented a paradox in this regard, in that the winning speeds for thoroughbreds, the result of three to four centuries' worth of ongoing selective breeding, appear not to have increased since the 1970s despite estimates of ample heritabilities for running performance. However, a study by Patrick

Sharman and Alistair Wilson of the University of Exeter expanded the scope of consideration to include speeds of both winners and losers over a range of distances. Doing so shows that racehorses in general are indeed still getting incrementally faster, but only for short-distance sprints, whereas middle- and long-distance speeds have plateaued. Thus, it may be that genetic variation exists primarily for those shorter-distance events but has been exhausted for longer-distance running.

Strong and unvarying selection over many generations can have other consequences beyond the depletion of genetic variation. Living organisms are tremendously complex, and the individual traits that organisms express do not exist in isolation of others. Traits such as performance are the emergent patterns of an intricate genetic tapestry of interwoven causal factors, from morphology and physiology to behavior. Change any one factor, and you could unwittingly change one or several others by dint of the close relations among them. These linkages exist because most genes influence several traits at the same time and because groups of different genes that influence a given trait are sometimes inherited together. This shared function means that those traits share genetic variation between them. In other words, a certain amount of a trait's additive genetic variation is shared with one or more other traits, and that shared variation is referred to as *genetic covariance*.

When enough genes influence two traits in the same way, such that the expression of a particular phenotype directly influences the expression of another through their shared genetic underpinnings, we can measure the strength of the relations among those traits as a number called the *genetic correlation*. These correlations may be positive or negative, and they form the heritable basis of specific genetic combinations of traits, such as the fast-developing, jumpy, but unattractive male crickets that I discussed in Chapter 3 (and also ultimately determine the patterns of life history trade-offs that I discussed in Chapter 8). Certain complex morphologies in particular depend heavily on large numbers of tight, closely integrated genetic correlations among traits to maintain functional coherence across generations. For example, snakes must swallow large food items without the benefit of limbs. The notion that snakes unhook their jaws is a myth; instead, the lower mandible is attached to the skull via a bone, called the quadrate, that pivots freely and allows the jaws large flexibility of movement. Snakes' skulls and lower jaws are therefore superbly adapted to accommodate large prey items, with floating joints, stretchy tendons connecting two halves of the jaw at the chin, and a variety of dynamic elements to the skull that together allow the mouth (and ultimately the head) to expand over and encompass large prey items. All of the morphological elements making up snake feeding systems are inherited together (a phenomenon called *phenotypic integration*) that is facilitated by patterns of integrated genetic correlations.

Those patterns of genetic correlations among groups of traits greatly influence the combinations in which those traits can be expressed and can constrain the direction of evolution in two ways: by biasing selection to operate on particular combinations of traits that have the most genetic variation available (because a greater response to selection might be obtained for those traits than for any other trait combination), or by preventing certain trait combinations from ever existing. Impossible trait combinations are said to reside in areas of nearly null genetic subspace and impose an absolute constraint on trait evolution that can be overcome only by an infusion of novel genetic variation.

Sand crickets (*Gryllus firmus*) come in two types, or morphs: migratory long-winged morphs that rely on flight to disperse to other areas and nonmigratory short-winged morphs that do not. The decision as to whether a given individual develops into a long-winged or a short-winged cricket is under both genetic and environmental control, but the two morphs themselves differ in both the type of traits expressed and the genetic relations among those traits. Long-winged crickets, because they are adapted for dispersal via flight, have relatively larger flight muscles than short-winged males, in addition to their larger wings, but they also call less and have smaller testes, likely because limited resources cannot be invested optimally in reproduction (that is, growing testes and attracting lady crickets) and flight muscle at the same time. These relations, such as the negative correlation between flight muscle and testes size, are reflected at the genetic level in the long-winged males, such that it is impossible to be a long-winged male with both large, functional flight muscles *and* large testes. A similar genetic trade-off is manifest in females, and long-winged females cannot produce both large ovaries and large flight muscles, and for similar reasons. Genetic correlations can thus produce similar limitations on trait evolution to the mechanical constraints discussed in Chapter 7.

The tension between the nature and nurture aspects of traits means we cannot simply look at phenotypic relations among traits and suggest that they are indicative of the underlying genetic correlations; the direction and strength of genetic correlations can be masked by phenotypic plasticity, and thus those genetic correlations must be estimated from breeding designs that manipulate relatedness among individuals while holding environmental conditions constant. Unfortunately, because estimating these genetic parameters for whole-organism performance traits is difficult at best, and because most animals are not amenable to estimation of genetic variances under laboratory conditions, our knowledge of these genetic athletic constraints is limited. In principle, however, there is nothing about performance that suggests they should *not* be subject to such constraints, and we have every reason to believe that the evolutionary trajectories of animal athletic abilities will be modified in the same way as for any other trait.

Keeping It in the Family

The breeding and experimental selection approaches that I have described thus far all fall under the purview of *quantitative genetics*, a branch of genetics specifically designed to deal with continuous phenotypes where the number and identities of specific genes affecting expression of those phenotypes are unknown, for all of the reasons I described at the beginning of this chapter.

Sometimes referred to as a way of doing genetics without genes, because knowledge of the specific genes involved in phenotypic expression is not required, quantitative genetics uses phenotypic resemblance among relatives to figure out how much of that resemblance is due to genes and how much is accounted for by environmental influences. One way to think of this is to consider how certain characteristics run in families. This happens because families share a common gene pool, but individuals within families are related to one another to a greater or lesser extent. Siblings share, on average, roughly 50 percent of their genetic material with each other, and 50 percent with their parents, whereas the parents themselves are for all intents and purposes unrelated (feel free to insert your favorite jibe at stereotypical rural folk or House Lannister here). Half siblings, who share only one parent in common, have on average 25 percent genetic similarity, as do nieces and uncles, with the percentage of shared genes growing ever smaller as relations become more distant. Quantitative genetics takes advantage of these genetic similarities and differences across families to estimate statistically how much of the variation in an observed character is likely to be genetic—in other words, to estimate trait heritabilities, as well as the genetic correlations among multiple traits.

Although it is true that most performance traits are of the type that quantitative genetics studies have been designed to detect—that is, traits that are affected by many, many, many genes—researchers are nonetheless especially interested in knowing if there are any individual genes that have a disproportionate influence on performance and, if possible, to identify these *genes of large effect*. In the relatively short time since the inception of molecular biology and our ability to isolate specific DNA sequences, researchers working on human athletics have identified a variety of genes and alleles that are associated with specific phenotypes that bestow an advantage in athletic performance.

Some of these putative performance genes are better candidates than others. For example, a particular version of a gene called COL5A1 has been associated with less flexibility in collagen and thus in tendons. Because stiffer tendons store elastic energy better than super-stretchy ones (see Chapter 7 for why), the noted connection between this COL5A1 allele and running ability might be a result of greater energy storage capacity in tendons such as the Achilles tendon. Another gene called ACTN3 makes a particular kind of muscle protein called alpha-actinen 3 that is associated with fast sprint speed; however, like many other genes, ACTN3 comes in various flavors, and while having the wrong version of ACTN3 almost definitely means that you will never be a world-class sprinter, merely possessing the correct ACTN3 allele is not enough to make you super-fast. Yet another allele, a point mutation version of the EPOR gene, leads to the overproduction of red blood cells in the bearer. The Finnish Olympian Eero Mäntyranta carries this version of EPOR, as do several other members of the Mäntyranta family, and his 65 percent higher-than-normal hematocrit was likely a decisive factor in his haul of seven Olympic medals for cross-country skiing, perhaps the most demanding endurance sport of all.

The Athletic Genetic Lottery

Although knowledge of these putative performance genes is valuable, their discovery was a haphazard affair. The results of a very large collaborative project called the HERITAGE (HEalth, RIsk factors, exercise Training And GEnetics) Family Study encompassing universities in the United States and Canada have vastly increased our understanding of the genetics underlying human performance.

Masterminded by Claude Bouchard, now at the Pennington Biomedical Research Center at Louisiana State University, this remarkable study aimed to uncover the identity of the specific genes underlying human endurance performance. Claude and his many colleagues did this by measuring, over twenty weeks, both cycling endurance and the response to regular cycling training in 481 otherwise sedentary human beings belonging to ninety-nine two-generation families (that is, families encompassing both parents and offspring). Because this study was conducted across so many family units and included both close relatives and nonrelatives, the researchers could separate the nature and nurture components of the exercise response in the same way as is done for animal quantitative genetic breeding designs. The results of this study are remarkable, and it's worth taking some time to go through them in detail.

First, the HERITAGE team found that cycling endurance has an h^2 of 0.42, which means that genetic variation accounts for about 42 percent of the variation in cycling endurance capacity in sedentary humans, and this is roughly on par with h^2s for performance abilities in other animals. However, in this design the subjects were measured not just once for cycling ability but repeatedly—in other words, the subjects exercised. On average, the study participants increased their VO_2max by 19 percent over the course of the study. But despite all being subject to identical training regimes, individual responses varied, with 5 percent of participants showing little or no change and another 5 percent increasing their VO_2max 40 percent to 50 percent! Bouchard and his colleagues found that *the ability to respond to endurance training* exhibits genetic variation as well and that about 47 percent of the training response in terms of VO_2max was accounted for by additive genetic factors.

Surprisingly, these two kinds of variation are not genetically correlated with each other—that is to say, individuals who are predisposed to have exceptional endurance abilities are not necessarily also high responders for endurance training. In fact, based on this study, probably as few as one in every one thousand people are predisposed *both* to being exceptional endurance athletes *and* to improving their endurance abilities substantially through training. If we factor in the results of the study from Chapter 3 that showed that the very best male endurance cyclists are also considered to have the most attractive faces, we can infer that there is some probably even smaller proportion of the population who are genetically predisposed to having high VO_2max, responding maximally to endurance training, and being attractive. This means, among other things, that the universe is a staggeringly unfair place.

As if offering unexpected insight into the bleakness of the human condition weren't enough, Bouchard and his genetic coconspirators went further, applying a technique called *genome-wide association* (GWA). This involves taking DNA samples from all participants and sequencing those samples to reveal the genome of each individual in the study—that is, every piece of genetic information contained within each individual's DNA. By then asking if any particular alleles are associated more often with high endurance performance across everyone involved in the study, Bouchard's group identified roughly 120 candidate genes that are likely involved in determining endurance ability and the response to endurance training in humans. But although we can point to some of these genes and say, "These are probably important for endurance and exercise somehow," figuring out exactly what these genes do is another story.

In some cases, we do know: for example, the CaMK (calcium/calmodulin dependent protein kinase) family of genes promotes the expression of slow-twitch oxidative muscle proteins and is involved in increasing mitochondria density.[3] Similarly, the angiotensin-I–converting enzyme gene (ACE) has previously been associated with both endurance and strength depending on the version of ACE in question: the ACE-I version (which decreases the activity of the enzyme, leading to lower blood pressure and higher circulation) is associated with endurance, whereas the ACE-D version, which increases enzyme activity, has been linked to strength. Other genes, such as that coding for apolipoprotein E (APOE), have been harder to pin down in terms of performance.

Although we don't currently understand the role of all of these candidate genes, it is possible that we soon shall. One of the most valuable outcomes of the HERITAGE study is that it has given exercise researchers a road map of which genes to concentrate on in terms of determining endurance capacity. However, it is always worth bearing in mind the polygenic nature of performance, and it may be that some of these candidate genes, such as APOE, are a part of a complex, integrated network of genes that affect and even regulate one another, and thus influence endurance and the endurance response weakly and/or indirectly. There is growing discontent with the results of large GWA studies, and ever louder murmurings in the genetics literature that perhaps genes of large effect are less important than the collective contributions of the uncountable thousands of other genes that constitute important phenotypic and regulatory signal buried within the noise of DNA sequences. And as promising as some of these genes are, we must be careful, as always, not to stray too far down the road of genetic determinism—indeed, there is evidence that the human endurance training

3. Mitochondria are inherited through the maternal line because only eggs carry them, sperm having divested themselves of much of their intracellular machinery as they became smaller. Mitochondria also contain their own DNA (mtDNA), a legacy of their originally having been free-living cells that were incorporated into and became part of others, a phenomenon called *endosymbiosis*. The HERITAGE study turned up evidence that VO_2max is inherited primarily through the mother, suggesting a role for mitochondrial variation, and thus mtDNA, in determining VO_2max.

response itself is subject to GxE interactions, being more pronounced on a low-glycogen diet.

Transcribing Athletic Performance

The major lesson from the HERITAGE study is that there is a lot of phenotypic variation among humans in endurance performance, large and quantifiable proportions of which are attributable to genetic differences. When we turn to the nonhuman animal world, however, things are still not so easy.

Identifying specific genes and genetic variants and their probable link with particular human traits has become much easier, thanks to our recently acquired ability to sequence the human genome. Although whole-genome sequencing is becoming ever easier and cheaper, we are still a long way off from sequencing the genomes of every animal species that we might be interested in. But technological advances do allow us to probe the genetic codes of organisms and to link them with specific phenotypes in ways that we could never have done in the past. One such approach is thanks to the explosive growth of the field of *transcriptomics*, which involves sequencing the various kinds of RNA that decode the information encoded in DNA and turn it into the proteins that build phenotypes.

By identifying these transcription factors, which are often very short-lived and activated under specific conditions, we can see which metabolic and biochemical pathways (that is, cascades or sequences of physiological processes) are being turned on or off at any given time. One such key pathway is that regulated by hypoxia-inducible factor (HIF), which controls the development of oxygen delivery networks in animals from insects to humans. HIF senses intracellular oxygen and triggers physiological processes that enhance oxygen delivery under low-oxygen conditions, such as those experienced by animals at high elevations. This pathway is stimulated in humans that train at altitude, and HIF activation in insects prompts enhanced growth of the air-filled tubes called trachea that deliver oxygen directly to tissues and cells that require it.

James Marden of Pennsylvania State University and colleagues showed that the Glanville fritillary butterfly exhibits three alleles of a gene called

Sdhd that produces an enzyme named succinate dehydrogenase (SDH) in the mitochondria. SDH in turn regulates expression of the transcription factor HIF-1α via an intermediary molecule called succinate, which SDH breaks down. One of these alleles, *Sdhd M*, reduces SDH production, resulting in an accumulation of succinate. Because that succinate isn't removed by SDH, it increases in concentration until it binds to HIF1-α, stabilizing it and preventing it from degrading, which it would do otherwise. This increased level of HIF-1α stimulates tracheal growth and results in butterflies with two-times higher tracheal density in their flight muscles, which leads them to enhanced endurance flight performance under experimental low-oxygen environmental conditions.

The existence of these *Sdhd* alleles in natural Glanville fritillary populations suggests that these animals are in a good position to respond to selection should they find themselves under chronic low-oxygen conditions such as those imposed by higher elevations or increased aerobic activity, in which case individuals carrying the *Sdhd M* allele would enjoy a fitness advantage over those without the allele. The ability to identify transcription factors therefore allows Marden and colleagues to draw a straight line from gene to performance in this organism and to understand several of the steps in between. Transcriptomics approaches also grant us insight into the observed genetic variation in the exercise response. For example, a related transcription factor called vascular endothelial growth factor (VEGF) regulates the growth of new blood vessels in vertebrates; is activated within hours of a single bout of treadmill running in rats; and, in humans, appears to be activated to a greater extent in those with a larger exercise response, and not at all in nonresponders.

No Sex Please, We're Geckos

If all of this talk of genetic variation has you thinking about sex, then congratulations, you have the makings of an evolutionary biologist! The evolution of sex, whereby organisms reproduce by combining part of their genetic material with that of others, is, to this day, a complicated issue, and I'm characteristically going to skip over the complexities thereof because this book is long enough as it is. But one thing that we can (probably) all agree on is that through combining the sperm and eggs of males and females, respectively, during conception (and by mixing and matching the chromosomes inherited from each parent during the formation of sperm and eggs themselves), sexual reproduction generates new and occasionally unique gene combinations, thereby increasing genetic and phenotypic variation. This is hugely useful in that it provides opportunity for some new and potentially useful genotypic combination to come along, increasing the fitness of the lucky bearer relative to those with less useful combinations (although harmful new genotypes are also possible).

We can scale this up a bit more, though, and consider what happens when individuals from different populations (or even species) that have had little contact with one another for some time regain contact. Evidence suggests that if groups have been continuously isolated from each other for very long time periods, then they are unlikely to recognize each other as potential mates. Even if they do, their matings might well fail to be fruitful because the two groups have diverged genetically to such an extent that their reproductive and developmental machinery no longer meshes. Sometimes, however, those matings do give rise to viable offspring, and these offspring can differ from each of their parents in striking ways. One potential consequence of hybridization is polyploidy—an increase in the chromosome number.

Most animals carry two sets of chromosomes (one each from the mother and father) and are thus diploid. In certain cases, however, the hybrid offspring of two different species combine their respective chromosome counts—two species with ten chromosomes each, for example, may give rise to offspring with twenty chromosomes! Polyploidy is rampant in plants, less common in invertebrate animals, and least so in vertebrates.

In lizards, the most common consequence of polyploidy is sterility. Members of these hybrid species also tend to be exclusively female. But the lack of males does not always condemn hybrid lizards to extinction as one might expect, because those females in some hybrid species are nonetheless able to reproduce asexually, via a process known as *parthenogenesis*. Although these asexual females do not require sperm to fertilize their eggs, those of certain species nonetheless need to perform the act of sex to reproduce, engaging in pseudocopulations with other individuals that involve no transfer of genetic material but that nonetheless stimulate development and differentiation of the unfertilized ovum into an embryo that is a genetic clone of the mother. One such asexual lineage is the Australian gecko *Heteronotia binoei*. This all-female hybrid species is triploid, harboring three sets of chromosomes. Compared to their diploid progenitors, parthenogenetic *H. binoei* are not only more active at low T_bs (those from 10°C–15°C, or 50°F–59°F), but they also exhibit higher endurance capacities at those temperatures. They are not, however, better at everything: asexual *H. binoei*, to give but one example, are 150 times more likely to be infested with parasitic mites.

The reasons for this improved endurance ability of parthenogenetic geckos are unclear. One possible explanation is that hybrid species tend to be superior to their parental lineages because they enjoy increased genetic diversity. Simply put, melding two more-or-less independent gene pools results in novel genetic combinations, increasing the overall genetic variation of important fitness-related traits such as locomotion and giving natural selection more alleles to choose among.

The mechanisms underlying this *hybrid vigor* have some conceptual support from other animal species; for example, researchers at the University of Queensland found that hybridizing two closely related *Drosophila* species breaks up the patterns of genetic correlations in morphology and behavior that had been built up over generations within each species, freeing up genetic variation in regions of genetic subspace that had previously been inaccessible due to genetic constraints. By selecting for various trait combinations in their new, genetically liberated hybrid *Drosophila* flies, the researchers obtained measurable responses to selection on trait combinations that did not respond in either of the parent species.

Hybrid vigor does not, however, account for the existence of asexual hybrid species that show either unchanged or lessened locomotor ability relative to the parental lineages. For example, a comparison of five physiological traits (including speed, endurance, and exertion) among five species of parthenogenetic, nongecko, hybrid *Cnemidophorus* lizards and their six sexual parent species showed that performance in the asexual species was either unchanged (in the case of sprinting and exertion) or markedly worse (in the case of endurance) compared to that of the sexual lizards. A similar result was reported for swimming acceleration in *Phoxinus* fishes, with asexual hybrids performing worse than the parental sexual species.

It's hard to say exactly why performance increases in asexual geckos but decreases in asexual lineages from other taxa. An unusual group of salamanders may harbor a clue. Some species of asexual salamanders in the genus *Ambystoma* display an unusual behavior whereby they steal the male spermatophores from a related sexual species and use those to stimulate egg development but do not necessarily incorporate the male DNA into their eggs. Yet others astonishingly incorporate DNA from multiple species into their eggs (a process called kleptogenesis), making them a sort of patchwork hybrid composed of parts of genomes stolen from other species!

A study led by Robert Denton of Ohio State University using several *Ambystoma* species finds that the lone asexual species they considered has the worst locomotor endurance capacity, roughly four times lower than that of its sexually reproducing cousins. Denton and colleagues suggest that a mismatch between the mitochondria and other genetic components swiped from various species results in a lower ability to use oxygen efficiently in the asexual species, which is certainly plausible. However, yet another fascinating animal system suggests some performance influence of sex chromosomes in particular.

African pygmy mice have a similar genetic sex determination system to that of our own, with females having two X chromosomes (XX) and males having an X and a Y (XY). But these animals also carry a third sex chromosome, called X*. X*Y mice do not develop into males—they stay physically female—but they behave more like males than XX females. These X*Y females not only are more aggressive and have higher reproductive success than regular females but also have larger heads and bite forces that are higher even than those of males! As is often the case, animals such as the African pygmy mouse raise far more questions than they answer, and the existence of these super females tells us that we still have much to learn about the genetic basis of performance.

TEN

Mice and Men

No examination of athletic abilities in the animal world is complete without considering human beings. The popularity of sports and athletic pastimes means that humans as a group arguably exhibit and regularly use more different types of performance abilities than any other animal on the planet, and in the early twentieth century the physiologist A. V. Hill realized that records of sporting events represent a valuable source of data on human athletic abilities. Researchers in subsequent decades have taken advantage of such data, often in combination with laboratory studies, to learn a great deal about human performance abilities and their morphological and physiological underpinnings. Indeed, the relations between morphology and performance explored in Chapters 6 and 7 are better understood in humans than perhaps any other animal thanks to the efforts of sports science and exercise physiology labs around the world. It is therefore odd that despite this wealth of functional data on human performance, humans are seldom used to test ideas from ecology and evolutionary biology in the same way that animal subjects are, and studies on human performance are conducted, for the most part, with a very different mindset.

Humans are, of course, animals, and like all animals we are (and have been) subject to evolution by natural and sexual selection. But although I have drawn freely on examples of human performance throughout this book to illustrate evolutionary concepts, this inclusive view of animal performance is not shared by everyone. It is curious that researchers working on performance in humans and nonhumans have tended to conduct their research entirely independent of one another. The Venn diagram of human and animal performance researchers intersects only a little, and the result is two very large, distinct, and complementary bodies of knowledge on performance that emphasize entirely different aspects: the human performance and sports medicine literature concentrates on kinematics, energetics, biomechanics, and nutrition, and the animal performance literature emphasizes comparative mechanics, ecological context, physiological ecology, and fitness correlates across a huge number of differing animal species.

Consequently, work on human performance has been largely conducted outside of the evolutionary and ecological framework that animal performance researchers use to interpret such data in nonhumans (although, to be fair, most researchers working on human performance are not primarily or even secondarily interested in questions about evolutionary ecology). When human performance researchers do turn to animal models (usually other primates or mice) for insight, those models are thus used only as guides or resources for what they can tell us about humans. From the reciprocal perspective, data on humans are seldom incorporated into comparative animal studies where they may in certain cases be highly relevant.

In some ways, this reluctance may be the legacy of turf wars between evolutionary biologists and scientists in other disciplines who historically resented the intrusion of evolutionary thinking into areas of human behavior in particular that were thought to be somehow special or otherwise exempt from the influences of selection. Humans are not special, but one does not have to look far to find opposing views. Fortunately, there have always been individuals on both sides willing to bridge this divide. More and more, researchers are bringing tools from each realm of research to bear on questions in the other. In this final chapter, I will revisit several themes introduced earlier and highlight studies involving humans that illustrate the insights afforded by an integrative mechanistic and evolutionary approach both to human athletic ability and to evolutionary ecology.

Human Racing

Compared to other animals, most human performance abilities are distinguished only by their mediocrity. We don't run particularly fast or jump very far; we are unimpressive swimmers and divers; and we can't even fly under our own power. The fact of functional, evolutionary, and genetic constraints means that we can't be expected to be good at everything, but even so, our general lack of athletic prowess borders on embarrassing. Yet there is one athletic facility in which humans have no peers in the animal kingdom: long-distance endurance running.

Humans are alone among primates and one of only a very few mammals that can run for long distances in hot, arid environments without either overheating or succumbing to crippling fatigue. From the Tarahumara Indians, who famously run for hundreds of miles at a time through the canyons of northwestern Mexico, to the annual 32-km (22-mi.) Man versus Horse Marathon in Wales, where human runners compete (and occasionally win) against men on horseback, humans have proven time and again to be the ultimate endurance athletes. Such feats demand an explanation from evolutionary biology.

Archaeological and ethnographic evidence points to the emergence approximately two million years ago of a human foraging strategy known as persistence hunting, whereby early humans, just like modern-day wild dogs, would chase down prey at submaximal speeds over long distances until the prey animals slowed due to either fatigue or heat exhaustion, allowing the prey to be clubbed or speared (by humans, not by wild dogs). Extended bouts of long-distance running raise the core body temperature, and in animals that are covered in insulating fur and lose heat primarily through panting, this is a serious problem.

The risk of overheating in hot environments is very real for large mammals, which is yet another consequence of the square-cube law. Large animals have small surface areas relative to their volume, which puts less of that volume in contact with the external environment for purposes of heat exchange compared with smaller animals. This makes it harder for large animals to lose heat, whereas small animals do so readily, and is the physical basis of gigantothermy discussed previously. But whereas some gazelles have evolved adaptations to tolerate heat storage (see Chapter 5), humans instead exhibit a suite of traits that allow us to efficiently shed much of the heat that accumulates over long-distance runs. There is mounting evidence that many of the features that distinguish us from other primates—lack of fur, high density of sweat glands, enlarged gluteus maximus muscles that contract little while walking but that stabilize the center of mass during running, enlarged semicircular canals in the ear that improve sensitivity to rapid changes in pitch while running—are adaptations that enable us to excel at aerobic running over distances that would exhaust or even kill both our primate relatives and the type of large mammalian prey that human hunter-gatherers favor. Sweating in particular is so effective as a cooling strategy that it has been incorporated into the designs of new robots to prevent overheating.

The kinematics of human endurance running also work in our benefit. For quadrupeds, there exists a speed above which the animal necessarily transitions from trotting to galloping. As bipeds, we lack the ability to gallop (for which two more legs are required), but our preferred endurance running speeds are above the trot-gallop transition speeds for our quadruped prey. Thus, humans can run comfortably for long distances at speeds where quadrupeds are forced to gallop, which is expensive and unsustainable for them over long distances. Quadrupeds also cannot employ their major method of heat loss, panting, while galloping. The end result is that we can run our prey into the ground as they succumb over time and distance to the combined effects of fatigue and overheating.

Proposed in 1984 by David Carrier and later championed by David Bramble of the University of Utah and Daniel Lieberman of Harvard University, the notion of humans as persistence hunters is unpopular in certain circles. Critics especially like to point out that persistence hunting is rare in contemporary indigenous human groups. However, it is important to recognize the historical and environmental context in which persistence hunting likely evolved and exists today. Population densities of potential prey today are substantially lower than was the case in even our recent history, making suitable prey harder to locate. Indigenous persistence hunters also currently lack the freedom of movement that they once enjoyed, and extensive fencing in places like Botswana and elsewhere in southern Africa further constrain their movement. What is more, persistence hunter-gatherers can obtain sufficient food by hunting only two to three days a week; consequently, there is no need for individuals practicing this strategy to be constantly hunting. (One anthropologist also makes the wry observation that modern researchers with sedentary lifestyles are unlikely to be accurately quantifying the hunting frequencies of persistence hunters, which would involve joining them in running down antelopes!)

Finally, persistence hunting may have been more recently supplanted by strategies based on the invention of the bow and arrow and the domestication of dogs and horses. But even though the conditions that led to the evolution of persistence hunting may no longer exist, as for pronghorns the evolutionary legacy of this superlative strategy remains. Strange as it may seem, our proclivity for competing in marathons may be the ultimate expression of what it is to be human.

The Apes of Wrath

David Carrier is not afraid to court controversy with his ideas regarding human evolution. More important, he can back up those ideas with evidence. Another of Carrier's propositions has proved to be even more controversial than persistence hunting: that the evolution of the human hand occurred at least partly in response to sexual selection for fighting ability.

The genesis of this idea has its roots in a paper in which Carrier and colleagues suggested that the sinking of whaling ships by sperm whales is evidence of a male combat strategy based on head-butting. In 1821, the American whaling ship *Essex* was sunk by a male sperm whale that used its head, buttressed by the characteristically large and oil-filled spermaceti organ (so called because of the semenlike texture of the oil inside), as a battering ram to break through the hull even though the *Essex* was much larger, heavier, and stronger than the whale. Traditional biological explanations for the spermaceti organ have focused on biosonar and buoyancy control, but Carrier and colleagues proposed instead that the spermaceti organ evolved as a weapon for male combat, which explains why the bull whale could use it so effectively to sink a ship. Male sperm whales have been observed head-butting in just this way, and the spermaceti organ itself is much larger in males compared to females, just as one might expect from something that is used predominantly in male combat. Head-butting has also been reported in other cetaceans such as dolphins and porpoises, and a head region equivalent to the spermaceti organ called the melon is found in other cetaceans. Male-biased sexual dimorphism is linked to melon size across twenty-one cetacean species, from killer whales and dolphins to narwhals, such that the males of more dimorphic species have larger melons than the females—again, a pattern consistent with that of a male weapon.

The researchers modeled the impact of an attacking bull whale of 39,000 kg (86,000 lb.), with a spermaceti organ that made up 20 percent of its body mass (7,800 kg, or around four and a half times the weight of a small car), moving at 3 m/s (~10.7 kph, or 6.7 mph), the estimated sum of the velocities of the *Essex* and the whale. They showed that the accelerations presented in this scenario, which were clearly enough to breach a ship's hull, are also at the level that would probably injure any other whale that finds itself the target of a male sperm whale headbutt while leaving the attacking whale uninjured, thanks to the spermaceti organ's shock-absorbing qualities. Given that sperm whales are far from ideal laboratory subjects, modeling studies might be as close as we will ever get to measuring the performance of whales during male combat, and the evidence, though circumstantial, suggests that the spermaceti organ could well have evolved to function in this context.

Not all were convinced by the whale study. One biologist in particular made a forceful argument in person to Carrier against the spermaceti-as-weapon hypothesis, brandishing his fist and saying, "I could hit you in the face with this, but it did not evolve for that!"

This gave Carrier an idea: What if human fists evolved as weapons for hitting others in the face? To test this hypothesis, Michael Morgan and Carrier measured the forces experienced by hands during open-hand and closed-fist striking and found that the bones of the human hand are proportioned in a way that buttresses and protects the hand from injury when striking forcefully with a clenched fist. This characteristic evolved early in our hominin lineage and is not present in any of our primate relatives, implying that fist-based male combat has historically been important in human evolution.

The suggestion that fists evolved for purposes of male combat drew immediate criticism. Accusations of adaptive storytelling were leveled, with some researchers adopting a position similar to that held by critics of the spermaceti organ hypothesis. Fresh attacks also arose from other researchers with experience in martial arts. Real fights between humans are of course nothing like they are represented in martial arts movies, being in reality inelegant, scrappy affairs, and older action movies are more accurate in this regard.[1] But one aspect of human fights that movies seldom address is the probability of severe hand injury when striking a bony human head with a closed, unprotected fist, natural buttressing or no. Padded gloves do a much better job of protecting the hands of punchers than they do the heads of punchees, and certain traditional martial arts styles promote an open-handed slap in preference to a closed-fist punch for this reason. Thus, anyone proposing an evolutionary scenario in which hands are used frequently as weapons must show not only that they are protected against damage when used but also that fistfights are the predominant way by which our ancestors resolved escalated conflicts despite the high risk of self-injury. In the same vein, yet others pointed out that if fists did evolve as weapons, then there should have been coevolution between fists and facial morphology to protect the area that fists struck at more often.

These objections led directly to Carrier and Morgan's next paper on the subject, in which they showed that fistfights are likely to be the predominant way by which our ancestors resolved escalated conflicts and that there has been coevolution between fists and facial morphology to protect the area that fists strike more often. To gain insight into the primary targets of assault in fights between untrained humans, they turned to epidemiological injury data. Not only is the face the most common site of injury in victims of interpersonal and domestic violence, with 81 percent of domestic violence victims in the United States sustaining injuries to the face and the middle part of the face being injured 69 percent of the time, studies on injury rates in the United Kingdom, Denmark, and Sweden refute the criticism of fists as being too fragile to be used in combat. Between 46 percent and 67 percent of facial fractures in those studies were caused by fists, yet associated fractures of the metacarpal or phalangeal bones in the hand were rare. To

1. Sometimes to a fault. The fight scenes in the otherwise underrated 1969 James Bond movie *On Her Majesty's Secret Service*, for example, look as if they were choreographed by an angry seal. George Lazenby in particular appears incapable of throwing a punch without falling over.

quote Carrier and Morgan directly: "Thus, human fists are common and effective weapons, and when humans fight, faces break much more frequently than fists."

Turning to facial morphology in humans and our hominin ancestors, Carrier and Morgan then asked whether there is any evidence of protective buttressing in those facial bones that are most at risk of being broken during a contentious interpersonal encounter with another human. There is. In our australopithecine ancestors, those bones are extremely robust, and much more so in males relative to females. The large jaw adductor muscles would also have served both as shock absorbers and jaw stabilizers when taking a punch, and the large postcanine teeth could have transferred punch energy from the jaw to the rest of the skull.

This pugilism hypothesis regarding human facial structure is supported by other evidence. Modern men with higher levels of circulating testosterone have broader and more robust faces than lower-testosterone men. Because of the influence of testosterone on strength and muscle function, one might expect that men with broader faces and more testosterone might also be better fighters, and this proves to be the case: fight outcomes between professional fighters can be predicted on facial broadness (that is, facial width-to-height ratio, or fWHR) alone, with broader-faced fighters being more likely to win fights. What is more, a separate study found that women, when presented with pictures of pairs of fighters (that is, a winner and a loser of a given bout, with winner-loser status unbeknown to the women), both predicted the winners more often than chance and rated the winners as more attractive than the losers. If exceptional cyclists are also considered to be sexy, as the work on facial cues discussed back in Chapter 3 shows, and men with broad faces are good fighters, it makes me wonder what range of fWHR is present in elite endurance athletes as opposed to those who have risen to the top ranks of combat sports.

Sinister Southpaw Sporting Strategies

In addition to showing that humans are prone to the same kinds of selection pressures that shape fighting ability in other animal species, sports studies can offer insight into aspects of male combat seen in other organisms. Left-right asymmetries are common throughout the animal kingdom and are often manifest as the phenomenon of handedness, whereby organisms show a preference or bias in terms of using either left or right limbs to perform certain tasks. Handedness is found in both vertebrates and invertebrates, but it is especially well described in humans. Around 90 percent of humans are right-handed, with the remaining 10 percent or so (including me) being left-handed, showing cross-dominance or mixed-handedness (doing some things left-handed and other things right-handed), or, in rare cases, being able to perform tasks equally well with both.

Because of the overwhelming prevalence of right-handers, most devices that require single-hand operation are designed for the use and benefit of right-handed people. Left-handers face the daily struggle of performing simple tasks like using scissors, corkscrews, and can openers without seriously injuring ourselves or others. There are other downsides as well—since the 1980s a spate of studies has shown that, compared to righties, left-handers are particularly prone to such hilarious scenarios as developing schizophrenia, having an autoimmune disorder, and dying on average nine years sooner, quite possibly through misadventure involving a chain saw. A 2004 study addressing the curious paucity of left-handed surgeons even found that 10 percent of such surgeons expressed concern at being treated themselves by another left-handed surgeon and provided a short but unsettling list of surgical instruments that left-handed surgeons felt "considerable difficulty" in handling.

These issues even extend to language. The Latin root of the word "dextrous" is *dexter*, which means right-oriented, whereas the original word for left-oriented is *sinister*, which today is a common synonym for "evil." Someone who can do things exceptionally well with left or right hands is described as "ambidextrous"—literally, "two right hands"—whereas a person clumsy with both is considered "ambisinister"! What is more, the French word for "left" is *gauche*, which in English means "awkward" or "inelegant." The indignities go on, but the upshot is that sinistral individuals such as myself exist in a hostile and uncaring dextral world that appears to be actively trying to murder us.

After a few decades of constantly sucking at simple tasks through no fault of your own and then discovering that you won't have to deal with it much longer because you will soon be dead anyway, it's easy to feel hard done by. However, there may be an upside to left-handedness that makes up for these shortcomings. Given that equal development and use of traits on both sides is the "default" evolutionary setting for organisms like us that express the developmental program of bilateral symmetry, deviations from such symmetry, especially genetically influenced ones where one side is favored consistently over the other, are likely the result of selection acting either directly on that asymmetrical trait or indirectly via a genetic correlation with something else that is being selected. This means that there is likely a fitness advantage associated with directional selection for handedness. Left-handedness in humans is both rare and heritable (with a broad-sense heritability of ~0.3), which gives us a clue as to one type of selective benefit that might accrue to southpaws.

The concept of negative frequency-dependent selection touched on briefly in the section on fiddler crabs in Chapter 3 describes situations where rare strategies or traits enjoy a competitive advantage over common ones. The unimaginatively named *fighting hypothesis* posits that left-handers, because of their rarity, realize an advantage in fights with right-handers that is sufficient to explain the low but persistent frequency of non-right-handed individuals in human populations. The rationale underlying a rare-handed advantage is that right-handed combatants facing lefties must adjust their fighting style to deal with attacks from unfamiliar angles, which puts them at a subtle disadvantage, especially in fights involving single-handed weapons, whereas left-handers who regularly fight righties have no such handicap. In light of this, the very best strategy (adopted by both Inigo Montoya and the Dread Pirate Roberts in *The Princess Bride*) is to be proficient in left- *and* right-handed fighting; however, this is far easier said than done. Although intuitive, the fighting hypothesis's negative frequency-dependence is not the only potential explanation for a southpaw advantage; for instance, another is that left-handers might have overall better spatial and visual skills thanks to their enlarged right brain hemisphere regions.

The fighting hypothesis has some empirical support that could potentially also explain the cultural anti-sinister bias. Homicide rates are positively correlated with the frequency of left-handers across several traditional human societies, with left-handers making up around 3 percent of the population in pacifist societies but accounting for as much as 27 percent of the population in those societies with more bellicose tendencies. This is suggestive of a combat-based advantage to left-handedness, albeit indirect. But we might also be able to gain some insight into the competitive benefits of left-handedness from human sporting events, including but not limited to combat sports.

Left-handers are overrepresented in interactive sports such as fencing and tennis but are not disproportionately common in noninteractive sports like gymnastics. High frequencies of left-handers even extend to interactive team sports. A 2004 study led by Rob Brooks of the University of New South Wales based on analyzing the results of the 2003 Cricket World Cup showed that left-handed batsmen were disproportionately overrepresented in the top world cricket teams. Such batsmen also performed better against the predominantly right-handed bowlers, especially those from lower-performing teams that had less experience bowling against left-handers. With regard to fighting specifically, left-handers are also overrepresented in the Ultimate Fighting Championship (UFC) yet interestingly don't win more fights than right-handers.

In an attempt to overcome some of the problems associated with earlier cross-sectional snapshot studies examining fight data at only one point in time, such as low sample size and the possibility that the dynamic relative frequencies of left- and right-handers might cause left-handers to cede their advantage as they become more common, researchers from the University of Kassel in Germany analyzed a large sample of boxers from 1924 to 2012. They found that southpaw boxers over time did indeed win more fights than they lost, providing direct support from a combat sport for the negative frequency-dependent nature of the fighting hypothesis.

Left-right limb preferences are especially strong in humans, but we are not the only animals to which the fighting hypothesis might apply. Fiddler crabs also show handedness, having either the right or the left claw enlarged into the major claw used for signaling and combat. But unlike in humans, handedness in most fiddler crab species is primarily plastic, showing a one-to-one ratio of left versus right major claws, and thus the influence of selection on left-right claw asymmetry is likely to be minimal. Indeed, tests of the fighting hypothesis in fiddler crabs have yielded inconsistent results, showing that the handedness of combatants does affect the way they fight when pitted against same- or different-handed opponents but has no effect on fight outcomes. However, at least 5 of the 102 described species of fiddler crabs are predominantly right-clawed, suggesting that true selection-driven handedness has evolved in those species in particular. Tests of the fighting hypothesis in those species might be especially illuminating.

Spectator Science

Human sports datasets have other advantages besides raw sample size in terms of testing ideas drawn from ecology and evolution. Professional athletes performing at the highest levels are likely to be very similar in terms of training levels, motivation, diet, and various other potentially confounding environmental factors, and that similarity facilitates performance comparison. Many sports also draw both male and female competitors at the highest levels, affording us insight into various athletic differences between the sexes.

The longitudinal nature of some sports, whereby we can follow individual athletes' performance over their careers, is another plus. As noted in Chapter 8, studies of aging are often hampered by the long lifespans of many organisms, which makes understanding how individuals alter their trait expression over time a challenge. In theory, human sports datasets should allow us to overcome this hurdle, but in practice things are not so easy. Whether thanks to age-related performance senescence or through injury-driven forced retirement, athletes' careers tend to be short compared to their overall lifespans. For every Kareem Abdul-Jabbar, who played in the National Basketball Association from the ages of twenty-two to forty-two, or National Hockey League Hall of Famer Ron Francis, whose career spanned twenty-three seasons from 1981 to 2004, there are many more athletes whose professional careers last only a year or two in their early twenties. Over a large enough timescale, however, we should be able to cobble together enough individual careers in particular sports to allow us to test some hypotheses.

A recurring theme throughout this book is that animals are subject to trade-offs in trait expression. But individuals do not necessarily roll over and accept the accompanying costs, and they often strive to compensate for them in some other way. We have already encountered several examples of this *trait compensation,* such as the female stalk-eyed flies in Chapter 4 that deal with the consequences of large eyestalks thrust upon them through intralocus sexual conflict by evolving different wing morphology to males. Studies have examined the contexts in which trait compensation is likely to exist, and they suggest that life history is an important factor; for example, brown anoles that exhibit greater exploratory behavior in novel environments compensate for the risks inherent in being a trailblazer by dropping their tails much more readily than less adventurous lizards, but only if food resources are abundant (probably because regrowing a lost tail isn't cheap). From an aging perspective, and since not all performance traits age at the same rate (see Chapter 8), one question of interest is whether animals compensate for senescence in particular performance abilities by increasing investment in others, thus staving off (at least temporarily) an overall decrease in fitness. This question is difficult to answer in nonhuman animals, but we can address it using human sporting data.

Scoring points in artificial physical competitions doesn't necessarily lead to higher individual fitness — at least, I am not aware of any studies that show athletic superstars to have higher fitness than regular athletes or any that test whether athletes exhibit higher fitness than nonathletes. But it can be considered to be the equivalent of succeeding in an ecologically relevant performance task. Scoring points is also the clearest metric of performance ability, even in team sports where all players are required to score for a team to be successful. In basketball, the three-point field goal, given for successful throws from outside of a designated three-point line surrounding the basket, had a contentious history since its inception in 1945 but was finally adopted by the NBA in the 1979–1980 season. This brought the number of ways to score in basketball up to three, including the regular two-point field goal awarded from within the three-point line and the one-point free-throw opportunities given to players fouled by the opposite team.

Two-pointers, three-pointers, and free throws require players to integrate different neuromuscular and performance attributes, such as jumping, accuracy, strength, and motor control, in different ways or to various extents. Only two-pointers can involve dunking, though, and since earlier studies showed that the peak age for Olympic long- and high-jump events over an eighty-six-year period were about twenty-two and twenty-four, respectively, one possibility is that players might compensate for reduced jumping ability (and thus potentially fewer two-pointers) as they age by increasing their scoring of three-point field goals. Furthermore, although the Women's NBA was founded only in April 1996, the rules of play are the same, and comparing aging trajectories between the NBA and WNBA allows us to test whether similar patterns of compensation (if any) are seen in men and women.

Analyzing all basketball scoring data from every NBA game from 1979 to 2010 to determine aging trajectories (while accounting for factors such as height, playing time, shooting success rate, and overall team quality) reveals that the peak scoring age for NBA players is twenty-five or slightly after (fig. 10.1a), beyond which there is a reasonably steep drop-off,[2] consistent with trends for many other individual sports, including those involving jumping. When the points scored are separated out by free throws, two-pointers, and three-pointers, the peak age for both free throws and two-pointers is, again, around twenty-five. However, the peak age for three-pointers occurs much later, at around thirty.

This appears to offer some support for the compensation hypothesis, and it would be easy to concoct a scenario wherein players begin increasing their accuracy at roughly the same time that their high-intensity, short-duration burst performance abilities are on the wane. However, given that the aging curve for overall points scored resembles much more closely those for two-point field goals and free throws, it is reasonable to infer that it is those methods of scoring rather than three-pointers that drive overall basketball

2. It is worth emphasizing that this represents the average age trajectory for 1,035 players over a thirty-one-year period. Michael Jordan, LeBron James, and everyone else who never seemed to decline in performance are all in there, as part of the variation delimited by the confidence intervals; but *on average*, the peak scoring age is twenty-five.

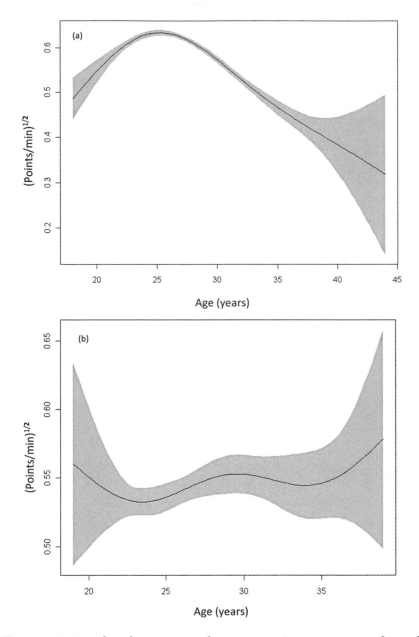

Fig. 10.1. Aging of total points scored per minute (square-root transformed for boring statistical reasons) in professional male (a) and female (b) basketball players. The shaded regions denote 95 percent confidence intervals, which illustrate the range of values within which the true slopes are likely to be found for each. Redrawn from Lailvaux et al. 2014.

scoring performance over time, and that any compensatory effect of three-pointers (if it exists) probably has little impact on overall points scored.

The second point of interest to emerge from the analysis of aging trends in professional basketball players is that WNBA players exhibit strikingly different aging trajectories to those seen in men both for overall points scored and for various score types. WNBA players show no clear senescence in overall scoring ability (fig. 10.1b). With regard to the specific means of scoring, only three-pointers show the familiar senescent pattern, reaching a peak at roughly age twenty-five that is followed by a long plateau. However, WNBA games have been played only since 1997–1998, and the truncated WNBA dataset (540 WNBA players over eleven years from 1998 to 2009 compared to 1,035 NBA players over thirty-one years) means that the results of the WNBA analysis are less robust.

Even despite those limitations, it nonetheless appears that scoring proficiency ages differently in men and women. These differences could be ascribed to any number of sources. For example, they may reflect innate physiological differences in the underlying performance abilities that drive scoring and lead to men and women playing the same game very differently, with dunking being rare in women's basketball thanks to the roughly 33 percent lower vertical leap ability of women compared with men. Yet another potential factor is the different scheduling and intensity of NBA and WNBA seasons. The regular NBA season, excluding playoffs, runs for eighty-two games, whereas a regular WNBA season comprises thirty-two. Even if NBA and WNBA players put in equivalent amounts of training, the average NBA player experiences around 140 percent more play time per season than the average WNBA player, and that more grueling play schedule alone could, over time, account for the occurrence of scoring senescence in male but not female professional basketballers.

Fallacious Footballers

That various sports involve differing kinds of player interactions and behavior has not escaped the attention of scientists interested in new ways of testing ideas from evolutionary theory. Matching a given sporting interaction to specific biological phenomena or predictions requires not only an intimate understanding of both the science and the sport at hand but also the perspicacity to connect the two in relevant ways, often during the weekend and after a couple of drinks. Here, as in most things sports related, Australia is leading the way. Researchers in the Performance Lab at the University of Queensland have turned the lens of evolutionary theory on aspects of soccer play and found fertile ground for evaluating some notions of honesty and deception.

As noted earlier, deception is challenging to test in nonhuman animals, given the difficulty in identifying signals that have evolved to mislead. Soccer players exhibit a notorious deceptive signaling behavior, diving, whereby intentionally falling mimics the behavioral consequence of an illegal tackle from an opposition player. Although a successful dive induces the signal receiver (the referee) to award the signaler (the diving player) a benefit in the form of a free kick, diving can also incur a receiver-imposed cost from the ref if the ruse is unsuccessful. Because diving has both benefits and potential costs, animal signaling theory suggests that it should occur most often when those benefits are maximized relative to the associated costs. For soccer matches specifically, this means that players should dive more frequently on areas of the pitch and during those parts of the game when doing so is likely to be most advantageous, and so diving should be both less common closer to the defensive goal and most common closest to the attacking goal. Similarly, the potential benefits of diving should be greatest when the score is even compared to when a team is currently winning or losing a match, because pulling ahead is of greater value than either equalizing or maintaining an existing lead.

Gwendolyn David watched sixty televised soccer matches (ten from each of six professional soccer leagues across the world) to test these predictions as part of her doctoral research. She used slow-motion replays and multiangle footage to categorize every player fall as either a dive, a tackle, or a fall for no signaling purpose (for example, if a player tripped or otherwise fell over in a nontackling context). She also divided the pitch up into zones and noted where on the pitch relative to the attacking and defensive goals the various falls took place, logged the score at the time of each fall, and estimated the proximity of each fall to the referee. Gwendolyn's results upheld these predictions, with players diving roughly twice as often in the zone closest to the attacking goal than in any of the defensive zones, and also diving more frequently when scores were tied as opposed to when the diver's team was either winning or losing.

Surprisingly, however, although referees are better able to detect deception the closer they are to a diving player, referees in this sample never punished an obvious dive. It is unclear why referees are reluctant to penalize clearly deceptive behavior, but explanations range from a predisposition to err on the side of leniency if there is any reasonable doubt to the arguably more popular opinion that some referees are worthless bastards. (Incidentally, soccer leagues in the United States and Australia that have begun to punish dives retroactively have seen a marked decrease in diving frequency.) Yet a further result to emerge from this study is that referees rewarded dives more often in leagues where diving was more prevalent, potentially explaining why this behavior is so common in players from those leagues. Gwendolyn and her coauthors declined to name the international league whose players dived most often, but we all know which it is.

Limitations of Spectator Science

Although analyses of cricket matches, mixed-martial-arts fights, basketball aging, and soccer diving demonstrate the scope for using existing sports data to test predictions from animal systems in novel and creative ways, they nonetheless also have their limits. Perhaps the biggest drawback is that the data are not derived from experiments designed to test specific hypotheses. Scientists always prefer to do experiments because if we can perform a manipulation and observe a subsequent change in the system that is not observed in an unmanipulated control, we can feel confident in inferring a causal relation between the thing that we manipulated and the thing that changed. If done correctly we can also estimate how much confidence we should place in that causal relation under those experimental conditions. But sports datasets are entirely ad hoc, and although we can in some cases look for situations that might mimic a manipulation (such as a change in the rules) and consider the consequences thereof, even that is less than ideal.

This ad hoc nature of sports datasets also creates problems in terms of data analysis. A key part of experimental design is planning manipulations so that the consequences thereof are as unambiguous as possible. Analysis of sports datasets often means controlling or accounting for a number of confounding variables and relations that would not exist in a properly designed experiment. Although some of those problems can be dealt with using appropriate statistical techniques, there are limits to the ways that one can (or should) brute force one's way through a complex and uncontrolled dataset without emerging with conclusions that are at worst wrong and at best misleading. But although using statistics to wring the informative, creamy center out of a rigid and fleshy dataset is not as satisfying as the more desirable coupling of experiment and inference, I nonetheless maintain that such endeavors have value. At the very least, they might generate new predictions that can serve as guides for more rigorous investigation by experiment at a later date.

Applying animal performance methods and ways of thinking to human athletic abilities brings us full circle. Despite our extensive cultural trappings, we have transcended neither the bounds of selection nor the reach of our evolutionary ancestry. The story of animal athletic performance is as much our own as it is that of the smasher shrimp or the flying snake. It follows, then, that to understand the nature of performance, from the smallest insect to the largest mammal, is to understand ourselves as well.

I have depended extensively on two truly excellent reference books for basic information on the mechanics and physiology of whole-organism performance: R. McNeill Alexander's *Principles of Animal Locomotion* and Steven Vogel's *Comparative Biomechanics: Life's Physical World* (both published by Princeton University Press). Each is notable for being well written and accessible to anyone with the time and interest to delve into them (although Alexander's is the more technical treatment). For more detailed case studies, I have relied on the published and peer-reviewed scientific literature, as well as the (very) occasional nonscientific resource. I list the primary references for each chapter below. Papers that provided information for multiple chapters are listed only under the first chapter for which they were a source.

Chapter One: Running, Jumping, and Biting

Alexander, R. McN., V. A. Langman, and A. S. Jayes. 1977. Fast locomotion of some African ungulates. *Journal of Zoology* 183: 219–300.

Brown, G. P., C. Shilton, B. L. Phillips, and R. Shine. 2007. Invasion, stress, and spinal arthritis in cane toads. *Proceedings of the National Academy of Sciences of the United States of America* 104: 17698–17700.

Christiansen, P., and S. Wroe. 2007. Bite forces and evolutionary adaptations to feeding ecology in carnivores. *Ecology* 88: 347–358.

Coburn, T. A. 2011. *The National Science Foundation: Under the Microscope*.

Dlugosz, E. M., et al. 2013. Phylogenetic analysis of mammalian oxygen consumption during exercise. *Journal of Experimental Biology* 47: 4712–4721.

Gilman, C. A., M. D. Bartlett, G. B. Gillis, and D. J. Irschick. 2012. Total recoil: perch compliance alters jumping performance and kinematics in green anole lizards (*Anolis carolinensis*). *Journal of Experimental Biology* 215: 220–226.

Lailvaux, S. P., et al. 2004. Performance capacity, fighting tactics and the evolution of life-stage male morphs in the green anole lizard (*Anolis carolinensis*). *Proceedings of the Royal Society of London B: Biological Sciences* 271: 2501–2508.

Llewelyn, J., et al. 2010. Locomotor performance in an invasive species: cane toads from the invasion front have greater endurance, but not speed, compared to conspecifics from a long-colonised area. *Oecologia* 162: 343–348.

Phillips, B. L., G. P. Brown, J. K. Webb, and R. Shine. 2006. Invasion and the evolution of speed in toads. *Nature* 39: 803.

Wilson, A. M., et al. 2013. Locomotion dynamics of hunting in wild cheetahs. *Nature* 498: 185–192.

Chapter Two: Eating and Not Being Eaten

Butler, M. A. 2005. Foraging mode of the chameleon, *Bradypodion pumilum*: a challenge to the sit-and-wait versus active foraging paradigm? *Biological Journal of the Linnean Society* 84: 797–808.

Crotty, T. L., and B. C. Jayne. 2015. Trade-offs between eating and moving: what happens to the locomotion of slender arboreal snakes when they eat big prey? *Biological Journal of the Linnean Society* 114: 446–458.

FitzGibbon, C. D., and J. H. Fanshawe. 1988. Stotting in Thomson's gazelles: an honest signal of condition. *Behavioral Ecology and Sociobiology* 23: 69–74.

Fu, S., et al. 2009. The behavioural, digestive and metabolic characteristics of fishes with different foraging strategies. *Journal of Experimental Biology* 212: 2296–2302.

Huey, R. B., and E. R. Pianka. 1981. Ecological consequences of foraging mode. *Ecology* 62: 991–999.

Kane, S. A., and M. Zamani. 2014. Falcons pursue prey using visual motion cues: new perspectives from animal-borne cameras. *Journal of Experimental Biology* 217: 225–234.

Leal, M., and J. A. Rodríguez-Robles. 1995. Antipredator responses of *Anolis cristatellus* (Sauria: Polychrotidae). *Copeia* 1995: 155–161.

Leong, T. M., and S. K. Foo. 2009. An encounter with the net-casting spider *Deinopis* species in Singapore (Araneae: Deinopidae). *Nature in Singapore* 2: 247–255.

Losos, J. B., T. W. Schoener, R. B. Langerhans, and D. A. Spiller. 2006. Rapid temporal reversal in predator-driven natural selection. *Science* 314: 1111.

McElroy, E. J., K. L. Hickey, and S. M. Reilly. 2008. The correlated evolution of biomechanics, gait and foraging mode in lizards. *Journal of Experimental Biology* 211: 1029–1040.

McHenry, M. J. 2012. When skeletons are geared for speed: the morphology, biomechanics, and energetics of rapid animal motion. *Integrative and Comparative Biology* 52: 588–596.

Mehta, R. S., and P. C. Wainwright. 2007. Raptorial jaws in the throat help moray eels swallow large prey. *Nature* 449: 79–83.

Miller, L. A., and A. Surlykke. 2001. How some insects detect and avoid being eaten by bats: tactics and countertactics of prey and predator. *BioScience* 51: 570–581.

Neutens, C., et al. 2014. Grasping convergent evolution in syngnathids: a unique tale of tails. *Journal of Anatomy* 224: 710–723.

Patek, S. N., and R. L. Caldwell. 2005. Extreme impact and cavitation forces of a biological hammer: strike forces of the peacock mantis shrimp *Odontodactylus scyllarus*. *Journal of Experimental Biology* 208: 3655–3664.

Patek, S. N., W. L. Korff, and R. L. Caldwell. 2004. Deadly strike mechanism of a mantis shrimp. *Nature* 428: 819–820.

Patek, S. N., J. E. Baio, B. L. Fisher, and A. V. Suarez. 2006. Multifunctionality and mechanical origins: ballistic jaw propulsion in trap-jaw ants. *Proceedings of the National Academy of Sciences of the United States of America* 103: 12787–12792.

Pruitt, J. N. 2010. Differential selection on sprint speed and *ad libitum* feeding behaviour in active vs. sit-and-wait foraging spiders. *Functional Ecology* 24: 392–399

Robinson, M. H., and B. Robinson. 1971. The predatory behavior of the ogre-faced spider *Dinopis longipes* F. Cambridge (Araneae: Dinopidae). *American Midland Naturalist* 85: 85–96.

Rose, T. A., A. J. Munn, D. Ramp, and P. B. Banks. 2006. Foot-thumping as an alarm signal in macropodoid marsupials: prevalence and hypotheses of function. *Mammal Review* 36: 281–291.

Van Wassenbergh, S. 2013. Kinematics of terrestrial capture of prey by the eel-catfish *Channallabes apus*. *Integrative and Comparative Biology* 53: 258–268.

Van Wassenbergh, S., G. Roos, and L. Ferry. 2011. An adaptive explanation for the horse-like shape of seahorses. *Nature Communications* 2:5.

Van Wassenbergh, S., et al. 2006. A catfish that can strike its prey on land. *Nature* 440: 881.

——. 2009. Suction is kid's play: extremely fast suction in newborn seahorses. *Biology Letters* 5: 200–203.

Westneat, M. W. 1991. Linkage biomechanics and evolution of the unique feeding mechanism of *Epibulus insidiator* (Labridae, Teleostei). *Journal of Experimental Biology* 159: 165–184.

Chapter Three: Lovers and Fighters

Allen, J. J., D. Akkaynak, A. K. Schnell, and R. T. Hanlon. 2017. Dramatic fighting by male cuttlefish for a female mate. *American Naturalist* 190: 144–151.

Andersson, M. 1994. *Sexual Selection*. Princeton, NJ: Princeton University Press.

Backwell, P. R. Y., M. D. Jennions, N. Passmore, and J. H. Christy. 1998. Synchronized courtship in fiddler crabs. *Nature* 391: 31–32.

Backwell, P. R. Y., et al. 2000. Dishonest signalling in a fiddler crab. *Proceedings of the Royal Society of London B: Biological Sciences* 267: 719–724.

Bely, A. E., and K. G. Nyberg. 2010. Evolution of animal regeneration: re-emergence of a field. *Trends in Ecology and Evolution* 25: 161–170.

Blackburn, D. C., J. Hanken, and F. A. Jenkins. 2008. Concealed weapons: erectile claws in African frogs. *Biology Letters* 4: 355–357.

Blows, M. W., R. Brooks, and P. G. Kraft. 2003. Exploring complex fitness surfaces: multiple ornamentation and polymorphism in male guppies. *Evolution* 57: 1622–1630.

Bywater, C. L., and R. S. Wilson. 2012. Is honesty the best policy? Testing signal reliability in fiddler crabs when receiver-dependent costs are high. *Functional Ecology* 26: 804–811.

Bywater, C. L., M. J. Angilletta, and R. S. Wilson. 2008. Weapon size is a reliable indicator of strength and social dominance in female slender crayfish (*Cherax dispar*). *Functional Ecology* 22: 311–316.

Emlen, D. J., J. Marangelo, B. Ball, and C. W. Cunningham. 2005. Diversity in the weapons of sexual selection: horn evolution in the beetle genus *Onthophagus* (Coleoptera: Scarabaeidae). *Evolution* 59: 1060–1084.

Hall, M. D., L. McLaren, R. C. Brooks, and S. P. Lailvaux. 2010. Interactions among performance capacities predict male combat outcomes in the field cricket *Teleogryllus commodus*. *Functional Ecology* 24: 159–164.

Husak, J. F., S. F. Fox, and R. A. Van Den Bussche. 2008. Faster male lizards are better defenders not sneakers. *Animal Behaviour* 75: 1725–1730.

Husak, J. F., A. K. Lappin, and R. A. Van Den Bussche. 2009. The fitness advantage of a high-performance weapon. *Biological Journal of the Linnean Society* 96: 840–845.

Husak, J. F., S. F. Fox, M. B. Lovern, and R. A. Van Den Bussche. 2006. Faster lizards sire more offspring: sexual selection on whole-animal performance. *Evolution* 60: 2122–2130.

Husak, J. F., A. K. Lappin, S. F. Fox, and J. A. Lemos-Espinal. 2006. Bite-force performance predicts dominance in male venerable collared lizards (*Crotaphytus antiquus*). *Copeia* 2006: 301–306.

Jacyniak, K. R., R. P. McDonald, and M. K. Vickaryous. 2017. Tail regeneration and other phenomena of wound healing and tissue restoration in lizards. *Journal of Experimental Biology* 220: 2858–2869.

Lailvaux, S. P., and D. J. Irschick. 2006. No evidence for female association with high-performance males in the green anole lizard, *Anolis carolinensis*. *Ethology* 112: 707–715.

——. 2007. The evolution of performance-based male fighting ability in Caribbean *Anolis* lizards. *American Naturalist* 170: 573–586.

Lailvaux, S. P., M. D. Hall, and R. C. Brooks. 2010. Performance is no proxy for genetic quality: trade-offs between locomotion, attractiveness, and life history in crickets. *Ecology* 91: 1530–1537.

Lailvaux, S. P., L. T. Reaney, and P. R. Y. Backwell. 2009. Dishonest signalling of fighting ability and multiple performance traits in the fiddler crab *Uca mjoebergi*. *Functional Ecology* 23: 359–366.

Lailvaux, S. P., J. Hathway, J. Pomfret, and R. J. Knell. 2005. Horn size predicts physical performance in the beetle *Euoniticellus intermedius*. *Functional Ecology* 19: 632–639.

Lappin, A. K., et al. 2006. Gaping displays reveal and amplify a mechanically based index of weapon performance. *American Naturalist* 168: 100–113.

Lee, S., S. Ditko, and A. Simek. 1963. Face to face with . . . the Lizard! *Amazing Spider-Man* 6.

McElroy, E. J., C. Marien, J. J. Meyers, and D. J. Irschick. 2007. Do displays send information about ornament structure and male quality in the ornate tree lizard, *Urosaurus ornatus? Ethology* 113: 1113–1122.

Meyers, J. J., D. J. Irschick, B. Vanhooydonck, and A. Herrel. 2006. Divergent roles for multiple sexual signals in a polygynous lizard. *Functional Ecology* 20: 709–716.

Mowles, S. L., P. A. Cotton, and M. Briffa. 2010. Whole-organism performance capacity predicts resource-holding potential in the hermit crab *Pagurus bernhardus*. *Animal Behaviour* 80: 277–282.

———. 2011. Flexing the abdominals: do bigger muscles make better fighters? *Biology Letters* 7: 358–360.

Nicoletto, P. F. 1993. Female sexual response to condition-dependent ornaments in the guppy, *Poecilia reticulata*. *Animal Behaviour* 46: 441–450.

———. 1995. Offspring quality and female choice in the guppy *Poecilia reticulata*. *Animal Behaviour* 49: 377–387.

Pomfret, J. C., and R. J. Knell. 2006. Sexual selection and horn allometry in the dung beetle *Euoniticellus intermedius*. *Animal Behaviour* 71: 567–576.

Postma, E. 2014. A relationship between attractiveness and performance in professional cyclists. *Biology Letters* 10: 20130966.

Reby, D., and K. McComb. 2003. Anatomical constraints generate honesty: acoustic cues to age and weight in the roars of red deer stags. *Animal Behaviour* 65: 519–530.

Snowberg, L. K., and C. W. Benkman. 2009. Mate choice based on a key ecological performance trait. *Journal of Evolutionary Biology* 22: 762–769.

Vanhooydonck, B., A. Y. Herrel, R. Van Damme, and D. J. Irschick. 2005. Does dewlap size predict male bite performance in Jamaican *Anolis* lizards? *Functional Ecology* 19: 38–42.

Wilson, R. S., et al. 2007. Dishonest signals of strength in male slender crayfish (*Cherax dispar*) during agonistic encounters. *American Naturalist* 170: 284–291

———. 2010. Females prefer athletes, males fear the disadvantaged: different signals used in female choice and male competition have varied consequences. *Proceedings of the Royal Society B: Biological Sciences* 277: 1923–1928.

Chapter Four: Girls and Boys

Andrade, M. C. B. 1996. Sexual selection for male sacrifice in the Australian redback spider. *Science* 271: 70–72.

Arnqvist, G., and L. Rowe. 2005. *Sexual Conflict.* Princeton, NJ: Princeton University Press.

Becker, E., S. Riechert, and F. Singer. 2005. Male induction of female quiescence/catalepsis during courtship in the spider, *Agelenopsis aperta. Behaviour* 142: 57–70.

Brodie, E. D. 1989. Behavioral modification as a means of reducing the cost of reproduction. *American Naturalist* 134: 225–238.

Cox, R. M., D. S. Stenquist, J. P. Henningsen, and R. Calsbeek. 2009. Manipulating testosterone to assess links between behavior, morphology, and performance in the brown anole *Anolis sagrei. Physiological and Biochemical Zoology* 82: 686–698.

Husak, J. F. 2006. Do female collared lizards change field use of maximal sprint speed capacity when gravid? *Oecologia* 150: 339–343.

Husak, J. F., and D. J. Irschick. 2009. Steroid use and human performance: lessons for integrative biologists. *Integrative and Comparative Biology* 49: 354–364.

Husak, J. F., G. Ribak, G. S. Wilkinson, and J. G. Swallow. 2011. Compensation for exaggerated eye stalks in stalk-eyed flies (Diopsidae). *Functional Ecology* 25: 608–616.

Husak, J. F., et al. 2013. Effects of ornamentation and phylogeny on the evolution of wing shape in stalk-eyed flies (Diopsidae). *Journal of Evolutionary Biology* 26: 1281–1293.

Huyghe, K., et al. 2009. Effects of testosterone on morphology, performance and muscle mass in a lizard. *Journal of Experimental Zoology Part A: Ecological Genetics and Physiology* 313A: 9–16.

Ketterson, E. D., V. Nolan, and M. Sandell. 2005. Testosterone in females: mediator of adaptive traits, constraint on sexual dimorphism, or both? *American Naturalist* 166: S85–S98.

Miles, D. B., R. Calsbeek, and B. Sinervo. 2007. Corticosterone, locomotor performance, and metabolism in side-blotched lizards (*Uta stansburiana*). *Hormones and Behavior* 51: 548–554.

Mohdin, A. 2015. Zoologger: oral sex may be a life saver for spider. *New Scientist* http://www.newscientist.com/article/dn26995-zoologger-oral-sex-may-be-a-life-saver-for-spider.html#.VOVrCmR4pCC.

Ramos, M., D. J. Irschick, and T. E. Christenson. 2004. Overcoming an evolutionary conflict: removal of a reproductive organ greatly increases locomotor performance. *Proceedings of the National Academy of Sciences* 101: 4883–4887.

Ramos, M., J. A. Coddington, T. E. Christenson, and D. J. Irschick. 2005. Have male and female genitalia coevolved? A phylogenetic analysis of genitalic morphology and sexual size dimorphism in web-building spiders (Araneae: Araneoidea). *Evolution* 59: 1989–1999.

Ribak, G., and J. G. Swallow. 2007. Free flight maneuvers of stalk-eyed flies: do eye-stalks affect aerial turning behavior? *Journal of Comparative Physiology A: Neuro-ethology, Sensory, Neural, and Behavioral Physiology* 193: 1065–1079.

Scales, J., and M. Butler. 2007. Are powerful females powerful enough? Acceleration in gravid green iguanas (*Iguana iguana*). *Integrative and Comparative Biology* 47: 285–294.

Seebacher, F., H. Guderley, R. M. Elsey, and P. L. Trosclair. 2003. Seasonal acclimatisation of muscle metabolic enzymes in a reptile (*Alligator mississippiensis*). *Journal of Experimental Biology* 206: 1193–1200.

Shine, R. 2003. Effects of pregnancy on locomotor performance: an experimental study on lizards. *Oecologia* 136: 450–456.

Swallow, J. G., G. S. Wilkinson, and J. H. Marden. 2000. Aerial performance of stalk-eyed flies that differ in eye span. *Journal of Comparative Physiology B* 170: 481–487.

Tarka, M., M. Åkesson, D. Hasselquist, and B. Hansson. 2014. Intralocus sexual conflict over wing length in a wild migratory bird. *American Naturalist* 183: 62–73.

Trivers, R. L. 1972. Parental investment and sexual selection. In *Sexual Selection and the Descent of Man*, 136–179. New York: Aldine de Gruyter.

Veasey, J. S., D. C. Houston, and N. B. Metcalfe. 2001. A hidden cost of reproduction: the trade-off between clutch size and escape take-off speed in female zebra finches. *Journal of Animal Ecology* 70: 20–24.

Warrener, A. G., K. L. Lewton, H. Pontzer, and D. E. Lieberman. 2015. A wider pelvis does not increase locomotor cost in humans, with implications for the evolution of childbirth. *PLoS One* 10: e0118903.

Webb, J. K. 2004. Pregnancy decreases swimming performance of female northern death adders (*Acanthophis praelongus*). *Copeia* 2004: 357–363.

Wells, C. L., and S. A. Plowman. 1983. Sexual differences in athletic performance: biological or behavioral? *Physician and Sportsmedicine* 11: 52–63.

Chapter Five: Hot and Cold

Alexander, R. M. 1989. *Dynamics of Dinosaurs and Other Extinct Giants*. New York: Columbia University Press.

Alonso, P. D., et al. 2004. The avian nature of the brain and inner ear of *Archaeopteryx*. *Nature* 430: 666–669.

Autumn, K., D. Jindrich, D. DeNardo, and R. Mueller. 1999. Locomotor performance at low temperature and the evolution of nocturnality in geckos. *Evolution* 53: 580–599.

Bakker, R. T. 1986. *The Dinosaur Heresies*. New York: Citadel Press.

Bennett, A. F., and J. A. Ruben. 1979. Endothermy and activity in vertebrates. *Science* 206: 649–654.

Chown, S. L., and S. W. Nicolson. 2004. *Insect Physiological Ecology: Mechanisms and Patterns*. Oxford: Oxford University Press.

Condon, C. H. L., and R. S. Wilson. 2006. Effect of thermal acclimation on female resistance to forced matings in the eastern mosquitofish. *Animal Behaviour* 72: 585–593.

Cowles, R. B. 1958. Possible origin of dermal temperature regulation. *Evolution* 12: 347–357.

Dial, K. P., B. E. Jackson, and P. Segre. 2008. A fundamental avian wing-stroke provides a new perspective on the evolution of flight. *Nature* 451: 985–983.

Else, P. L., and A. J. Hulbert. 1987. Evolution of mammalian endothermic metabolism: "leaky" membranes as a source of heat. *American Journal of Physiology: Regulatory, Integrative and Comparative Physiology* 253: R1–R7.

Feduccia, A. 1993. Evidence from claw geometry indicating arboreal habits of *Archaeopteryx*. *Science* 259: 790–793.

Gamble, T., et al. 2012. Repeated origin and loss of adhesive toepads in geckos. *PloS One* 7: e39429.

Gomes, F. R., C. R. Bevier, and C. A. Navas. 2002. Environmental and physiological factors influence antipredator behavior in *Scinax hiemalis* (Anura: Hylidae). *Copeia* 2002: 994–1005.

Grady, J. M., et al. 2014. Evidence for mesothermy in dinosaurs. *Science* 344: 1268–1272.

Gunn, D. L. 1933. The temperature and humidity relations of the cockroach (*Blatta orientalis*). *Journal of Experimental Biology* 10: 274–285.

Gunn, D. L., and C. A. Cosway. 1938. The temperature and humidity relations of the cockroach. *Journal of Experimental Biology* 15: 555–563.

Heinrich, B. 2013. *The Hot-Blooded Insects: Strategies and Mechanisms of Thermoregulation*. Berlin: Springer.

Herrel, A., R. S. James, and R. Van Damme. 2007. Fight versus flight: physiological basis for temperature-dependent behavioral shifts in lizards. *Journal of Experimental Biology* 210: 1762–1767.

Hertz, P. E., R. B. Huey, and E. Nevo. 1982. Fight versus flight: body temperature influences defensive responses of lizards. *Animal Behaviour* 30: 676–679.

Hetem, R., et al. 2013. Cheetah do not abandon hunts because they overheat. *Biology Letters* 9: 20130472.

Huey, R. B., and A. F. Bennett. 1987. Phylogenetic studies of coadaptation: preferred temperatures versus optimal performance temperatures of lizards. *Evolution* 41: 1098–1115.

Huey, R. B., and M. Slatkin. 1976. Costs and benefits of lizard thermoregulation. *Quarterly Review of Biology* 51: 363–384.

Huey, R. B., P. H. Niewiarowski, J. Kaufman, and J. C. Herron. 1989. Thermal biology of nocturnal ectotherms: is sprint performance of geckos maximal at low body temperatures? *Physiological Zoology* 62: 488–504.

Hulbert, A. J., and P. L. Else. 2000. Mechanisms underlying the cost of living in animals. *Annual Review of Physiology* 62: 207–235.

Kingsolver, J. G. 1985. Butterfly thermoregulation: organismic mechanisms and population consequences. *Journal of Research on the Lepidoptera* 24: 1–20.

Kramer, A. E. 1968. Motor patterns during flight and warm-up in Lepidoptera. *Journal of Experimental Biology* 48: 89–109.

Krogh, A., and E. Zeuthen. 1941. The mechanism of flight preparation in some insects. *Journal of Experimental Biology* 18: 1–10.

Lailvaux, S. P., G. J. Alexander, and M. J. Whiting. 2003. Sex-based differences and similarities in locomotor performance, thermal preferences, and escape behaviour in the lizard *Platysaurus intermedius wilhelmi*. *Physiological and Biochemical Zoology* 76: 511–521.

Martin, T. L., and R. B. Huey. 2008. Why "suboptimal" is optimal: Jensen's inequality and ectotherm thermal preferences. *American Naturalist* 171: E102–E118.

McNab, B. K. 2002. *The Physiological Ecology of Vertebrates: A View from Energetics*. Ithaca, NY: Comstock.

Pearson, O. P. 1954. Habits of the lizard *Liolaemus multiformis multiformis* at high altitudes in southern Peru. *Copeia* 1954: 111–116.

Pontzer, P., V. Allen, and J. R. Hutchinson. 2009. Biomechanics of running indicates endothermy in bipedal dinosaurs. *PLoS One* 4: e7783.

Ruben, J. 1991. Reptilian physiology and the flight capacity of *Archaeopteryx*. *Evolution* 45: 1–17.

Schaeffer, P. J., K. E. Conley, and S. L. Lindstedt. 1996. Structural correlates of speed and endurance in skeletal muscle: the rattlesnake tailshaker muscle. *Journal of Experimental Biology* 199: 351–358.

Seebacher, F., G. C. Grigg, and L. A. Beard. 1999. Crocodiles as dinosaurs: behavioural thermoregulation in very large ectotherms leads to high and stable body temperatures. *Journal of Experimental Biology* 202: 77–86.

Sellers, W. I., and P. L. Manning. 2007. Estimating dinosaur maximum running speeds using evolutionary robotics. *Proceedings of the Royal Society of London B: Biological Sciences* 274: 2711–2716.

Sellers, W. I., et al. 2017. Investigating the running abilities of *Tyrannosaurus rex* using stress-constrained multibody dynamic analysis. *PeerJ* 5: e3402.

Seymour, R. 2013. Maximal aerobic and anaerobic power generation in large crocodiles *versus* mammals: implications for dinosaur gigantothermy. *PLoS One* 8: e69361.

Shine, R., M. Wall, T. Langkilde, and R. T. Mason. 2005. Battle of the sexes: forcibly inseminating male garter snakes target courtship to more vulnerable females. *Animal Behaviour* 70: 1133–1140.

Shipman, P. 1998. *Taking Wing: Archaeopteryx and the Evolution of Bird Flight*. New York: Simon and Schuster.

Spotila, J. R., M. P. O'Connor, P. Dodson, and F. V. Paladino. 1991. Hot and cold running dinosaurs: body size, metabolism, and migration. *Modern Geology* 16: 203–227.

Taylor, C. R., and V. J. Rowntree. 1973. Temperature regulation and heat balance in running cheetahs: a strategy for sprinters? *American Journal of Physiology* 224: 848–851.

Wilson, R. S., C. H. L. Condon, and I. A. Johnston. 2007. Consequences of thermal acclimation for the mating behaviour and swimming performance of female mosquito fish. *Philosophical Transactions of the Royal Society B: Biological Sciences* 362: 2131–2139.

Chapter Six: Shape and Form

Arnold, S. J. 1983. Morphology, performance, and fitness. *American Zoology* 23: 347–361.

Biewener, A. A. 2003. *Animal Locomotion*. Oxford: Oxford University Press.

Blob, R. W., R. Rai, M. L. Julius, and H. L. Schoenfuss. 2006. Functional diversity in extreme environments: effects of locomotor style and substrate texture on the waterfall-climbing performance of Hawaiian gobiid fishes. *Journal of Zoology* 268: 315–324.

Bomphrey, R. J., T. Nakata, N. Philips, and S. M. Walker. 2017. Smart wing rotation and trailing-edge vortices enable high frequency mosquito flight. *Nature* 544: 92–95.

Bonine, K. E., and T. Garland. 1999. Sprint performance of phrynosomatid lizards, measured on a high-speed treadmill, correlates with hindlimb length. *Journal of Zoology* 248: 255–265.

Clifton, G. T., T. L. Hedrick, and A. A. Biewener. 2015. Western and Clark's grebes use novel strategies for running on water. *Journal of Experimental Biology* 218: 1235–1243.

D'Amore, D. C., K. Moreno, C. R. McHenry, and S. Wroe. 2011. The effects of biting and pulling on the forces generated during feeding in the Komodo dragon (*Varanus komodoensis*). *PLoS One* 6: e26226.

Davenport, J. 1994. How and why do flying fish fly? *Reviews in Fish Biology and Fisheries* 4: 184–214.

Dickinson, M. H., et al. 2000. How animals move: an integrative view. *Science* 288: 100–106.

Dudley, R., et al. 2007. Gliding and the functional origins of flight: biomechanical novelty or necessity? *Annual Review of Ecology, Evolution, and Systematics* 38: 179–201.

Flammang, B. E., A. Suvarnaraksha, J. Markiewicz, and D. Soares. 2016. Tetrapod-like pelvic girdle in a walking cavefish. *Scientific Reports* 6: 23711.

Fry, B. G., et al. 2009. A central role for venom in predation by *Varanus komodoensis* (Komodo dragon) and the extinct giant *Varanus (Megalania) priscus*. *Proceedings of the National Academy of Sciences* 106: 8969–8974.

Gilbert, C. 1997. Visual control of cursorial prey pursuit by tiger beetles (Cicindelidae). *Journal of Comparative Physiology* A 181: 217–230.

Glasheen, J. W., and T. A. McMahon. 1996a. Size-dependence of water-running ability in basilisk lizards (*Basiliscus basiliscus*). *Journal of Experimental Biology* 199: 2611–2618.

———. 1996b. A hydrodynamic model of locomotion in the basilisk lizard. *Nature* 380: 340–342.

Harpole, T. 2005. Falling with the falcon. *Air and Space Magazine* http://www.air spacemag.com/flight-today/falling-with-the-falcon-7491768/?no-ist=&page=1.

Hoyt, J. W. 1975. Hydrodynamic drag reduction due to fish slimes. In *Swimming and Flying in Nature*, vol. 2, ed. T. Wu, 653–672. Berlin: Springer.

Hudson, P. E., et al. 2011. Functional anatomy of the cheetah (*Acinonyx jubatus*). *Journal of Anatomy* 218: 375–385.

Humphrey, J. A. C. 1987. Fluid mechanic constraints on spider ballooning. *Oecologia* 73: 469–477.

Hutchinson, J. R., D. Famini, R. Lair, and R. Kram. 2003. Are fast-moving elephants really running? *Nature* 422: 493–494.

Jenkins, A. R. 1995. Morphometrics and flight performance of southern African peregrine and lanner falcons. *Journal of Avian Biology* 26: 49–58.

Johansson, F., M. Söderquist, and F. Bokma. 2009. Insect wing shape evolution: independent effects of migratory and mate guarding flight on dragonfly wings. *Biological Journal of the Linnean Society* 97: 362–372.

Krausman, P. R., and S. M. Morales. 2005. *Acinonyx jubatus*. *Mammalian Species* 771: 1–6.

Laybourne, R. C. 1974. Collision between a vulture and an aircraft at an altitude of 37,000 feet. *Wilson Bulletin* 86: 461–462.

Lentink, D., et al. 2007. How swifts control their glide performance with morphing wings. *Nature* 446: 1082–1085.

McGuire, J. A., and R. Dudley. 2005. The cost of living large: comparative gliding performance in flying lizards (Agamidae: *Draco*). *American Naturalist* 166: 93–106.

Miles, D. B., L. A. Fitzgerald, and H. L. Snell. 1995. Morphological correlates of locomotor performance in hatchling *Amblyrhynchus cristatus*. *Oecologia* 103: 261–264.

Myers, M. J., and K. Steudel. 1985. Effect of limb mass and its distribution on the energetic cost of running. *Journal of Experimental Biology* 116: 363–373.

Ropert-Coudert, Y., et al. 2004. Between air and water: the plunge dive of the Cape Gannet *Morus capensis*. *Ibis* 146: 281–290.

Sagong, W., W. Jeon, and H. Choi. 2013. Hydrodynamic characteristics of the sailfish (*Istiophorus platypterus*) and swordfish (*Xiphias gladius*). *PLoS One* 8: e81323.

Sharp, N. C. C. 2012. Animal athletes: a performance review. *Veterinary Record* 171: 87–94.

Socha, J. J. 2002. Gliding flight in the paradise tree snake. *Nature* 418: 603–604.

Socha, J. J, T. O'Dempsey, and M. LaBarbera. 2008. A 3-D kinematic analysis of gliding in a flying snake, *Chrysopelea paradisi*. *Journal of Experimental Biology* 208:
1817–1833.

Svendsen, M. B. S., et al. 2016. Maximum swimming speeds of sailfish and three other
large marine predatory fish species based on muscle contraction time and stride
length: a myth revisited. *Biology Open* 5: 1415–1419.

Van Valkenburgh, B., et al. 2004. Respiratory turbinates of canids and felids: a quantitative comparison. *Journal of Zoology* 264: 281–293.

Videler, J. J. 2006. *Avian Flight*. Oxford: Oxford University Press.

Videler, J. J., et al. 2016. Lubricating the swordfish head. *Journal of Experimental Biology*
219: 1953–1956.

Wang, L., et al. 2011. Why do woodpeckers resist head impact injury? A biomechanical
investigation. *PLoS One* 6: e26490.

Wassersug, R. J., et al. 2005. The behavioral responses of amphibians and reptiles to
microgravity on parabolic flights. *Zoology* 108: 107–120.

Wen, L., J. C. Weaver, and G. V. Lauder. 2014. Biomimetic shark skin: design, fabrica-
tion, and hydrodynamic function. *Journal of Experimental Biology* 217: 1656–1666.

Williams, T. M., et al. 1997. Skeletal muscle histology and biochemistry of an elite
sprinter, the African cheetah. *Journal of Comparative Physiology B* 167: 527–535.

Wojtusiak, J., E. J. Godzińska, and A. Dejean. 1995. Capture and retrieval of very large
prey by workers of the African weaver ant, *Oecophylla longinoda*. *Tropical Zoology*
8: 309–318.

Yafetto, L., et al. 2008. The fastest flights in nature: high-speed spore discharge mechanisms among fungi. *PLoS One* 3: e3237.

Young, J., et al. 2009. Details of insect wing design and deformation enhance aerodynamic function and flight efficiency. *Science* 325: 1549–1552.

Chapter Seven: Limits and Constraints

Abe, T., K. Kumagai, and W. F. Brechue. 2000. Fascicle length of leg muscles is greater
in sprinters than distance runners. *Medicine and Science in Sports and Exercise* 32:
1125–1129.

Alexander, R. M. 1991. It may be better to be a wimp. *Nature* 353: 696.

Bayley, T. G., G. P. Sutton, and M. Burrows. 2012. A buckling region in locust hindlegs
contains resilin and absorbs energy when jumping or kicking goes wrong. *Journal
of Experimental Biology* 215: 1151–1161.

Biewener, A. A. 2016. Locomotion as an emergent property of muscle contractile dynamics. *Journal of Experimental Biology* 218: 285–294.

Bro-Jørgensen, J. 2013. Evolution of sprint speed in African savannah herbivores in relation to predation. *Evolution* 67: 3371–3376.

Burrows, M. 2003. Froghopper insects leap to new heights. *Nature* 424: 509.

Burrows, M., S. R. Shaw, and G. P. Sutton. 2008. Resilin and chitinous cuticle form a composite structure for energy storage in jumping by froghopper insects. *BMC Biology* 6: 16.

Byers, J. A. 2003. *Built for Speed: A Year in the Life of Pronghorn.* Cambridge, MA: Harvard University Press.

Carrier, D. R. 2002. Functional trade-offs in specialization for fighting versus running. In *Topics in Functional and Ecological Vertebrate Morphology*, ed. P. Aerts, K. D'Août, A. Herrel, and R. Van Damme, 235–255. Maastricht: Shaker.

——. 1996. Ontogenetic limits on locomotor performance. *Physiological Zoology* 69: 467–488.

Costello, D. F. 1969. *The Prairie World.* New York: Thomas Y. Crowell.

Cullen, J. A., T. Maie, H. L. Schoenfuss, and R. W. Blob. 2013. Evolutionary novelty versus exaptation: oral kinematics in feeding versus climbing in the waterfall-climbing Hawaiian goby *Sicyopterus stimpsoni*. *PLoS One* 8: e53274.

Curry, J. W., R. Hohl, T. D. Noakes, and T. A. Kohn. 2012. High oxidative capacity and type IIx fibre content in springbok and fallow deer skeletal muscle suggest fast sprinters with a resistance to fatigue. *Journal of Experimental Biology* 215: 3997–4005.

Deban, S. M., and J. A. Scales. 2016. Dynamics and thermal sensitivity of ballistic and non-ballistic feeding in salamanders. *Journal of Experimental Biology* 219: 431–444.

de Groot, J. H., and J. L. van Leeuwen. 2004. Evidence for an elastic projection mechanism in the chameleon tongue. *Proceedings of the Royal Society B: Biological Sciences* 271: 761–770.

Farley, C. T. 1997. Maximum speed and mechanical power output in lizards. *Journal of Experimental Biology* 200: 2189–2195.

Hedenström, A., et al. 2016. Annual 10-month aerial life phase in the common swift *Apus apus*. *Current Biology* 26: 3066–3070.

Heers, A. M., and K. P. Dial. 2015. Wings versus legs in the avian *bauplan*: development and evolution of alternative locomotor strategies. *Evolution* 69: 305–320.

Hudson, P. E., et al. 2011. Functional anatomy of the cheetah (*Acinonyx jubatus*). *Journal of Anatomy* 218: 363–374.

Husak, J. F., and S. F. Fox. 2006. Field use of maximal sprint speed by collared lizards (*Crotaphytus collaris*): compensation and sexual selection. *Evolution* 60: 1888–1895.

Iosilevskii, G., and D. Weihs. 2008. Speed limits on swimming of fishes and cetaceans. *Journal of the Royal Society Interface* 5: 329–338.

Irschick, D. J., and J. B. Losos. 1999. Do lizards avoid habitats in which performance is submaximal? The relationship between sprinting capabilities and structural habitat use in Caribbean anoles. *American Naturalist* 154: 298–305.

Irschick, D. J., B. Vanhooydonck, A. Herrel, and A. Andronescu. 2003. The effects of loading and size on maximum power output and gait characteristics in geckos. *Journal of Experimental Biology* 206: 3923–3934.

Killen, S. S., J. J. H. Nati, and C. D. Suski. 2015. Vulnerability of individual fish to capture by trawling is influenced by capacity for anaerobic metabolism. *Proceedings of the Royal Society of London B: Biological Sciences* 282: 20150603.

Kohn, T. A., J. W. Curry, and T. D. Noakes. 2011. Black wildebeest skeletal muscle exhibits high oxidative capacity and a high proportion of type IIx fibres. *Journal of Experimental Biology* 214: 4041–4047.

Kropff, E., J. E. Carmichael, M. Moser, and E. I. Moser. 2015. Speed cells in the medial entorhinal cortex. *Nature* 523: 419–424.

Lindstedt, S. L., et al. 1991. Running energetics in the pronghorn antelope. *Nature* 353: 748–750.

Losos, J. B., and B. Sinervo. 1989. The effects of morphology and perch diameter on sprint performance of *Anolis* lizards. *Journal of Experimental Biology* 145: 23–30.

Marsh, R. L., and A. F. Bennett. 1986. Thermal dependence of sprint performance of the lizard *Sceloporus occidentalis*. *Journal of Experimental Biology* 126: 79–87.

McKean, T., and B. Walker. 1974. Comparison of selected cardiopulmonary parameters between the pronghorn and the goat. *Respiration Physiology* 21: 365–370.

Noakes, T. D. 2011. Time to move beyond a brainless exercise physiology: the evidence for complex regulation of human exercise performance. *Applied Physiology, Nutrition, and Metabolism* 36: 23–35.

Pasi, B. M., and D. R. Carrier. 2003. Functional trade-offs in the limb muscles of dogs selected for running vs. fighting. *Journal of Evolutionary Biology* 16: 324–332.

Quillin, K. J. 2000. Ontogenetic scaling of burrowing forces in the earthworm *Lumbricus terrestris*. *Journal of Experimental Biology* 203: 2757–2770.

Vanhooydonck, B., R. Van Damme, and P. Aerts. 2001. Speed and stamina trade-off in lacertid lizards. *Evolution* 55: 1040–1048.

Vanhooydonck, B., et al. 2014. Is the whole more than the sum of its parts? Evolutionary trade-offs between burst and sustained locomotion in lacertid lizards. *Proceedings of the Royal Society B: Biological Sciences* 281: 10.

Wainwright, P. C., M. E. Alfaro, D. I. Bolnick, and C. D. Hulsey. 2005. Many-to-one mapping of form to function: a general principle in organismal design? *Integrative and Comparative Biology* 45: 256–262.

Wakeling, J. M., and I. A. Johnston. 1998. Muscle power output limits fast-start performance in fish. *Journal of Experimental Biology* 201: 1505–1526.

Watanabe, Y. Y., et al. 2011. Poor flight performance in deep-diving cormorants. *Journal of Experimental Biology* 214: 412–421.

Weir, J. P., T. W. Beck, J. T. Cramer, and T. J. Housh. 2006. Is fatigue all in your head? A critical review of the central governor model. *British Journal of Sports Medicine* 40: 573–586.

Wilson, R. S., J. F. Husak, L. G. Halsey, and C. J. Clemente. 2015. Predicting the movement speeds of animals in natural environments. *Integrative and Comparative Biology* 55: 1125–1141.

Williams, S. B., et al. 2008. Functional anatomy and muscle moment arms of the pelvic limb of an elite sprinting athlete: the racing greyhound (*Canis familiaris*). *Journal of Anatomy* 213: 361–372.

Wolfman, M., and G. Pérez. 1985. A flash of lightning. *Crisis on Infinite Earths* 8. DC Comics.

Chapter Eight: Death and Taxes

Au, D., and D. Weihs. 1980. At high speeds dolphins save energy by leaping. *Nature* 284: 548–550.

Bailey, I., J. P. Myatt, and A. M. Wilson. 2013. Group hunting within the Carnivora: physiological, cognitive and environmental influences on strategy and cooperation. *Behavioral Ecology and Sociobiology* 67: 1–17.

Baker, A. B., and Y. Q. Tang. 2010. Aging performance for masters records in athletics, swimming, rowing, cycling, triathlon, and weightlifting. *Experimental Aging Research* 36: 453–477.

Biewener, A. A., D. D. Konieczynski, and R. V. Baudinette. 1998. In vivo muscle force-length behavior during steady-speed hopping in tammar wallabies. *Journal of Experimental Biology* 201: 1681–1694.

Bronikowski, A. M., T. J. Morgan, T. Garland, and P. A. Carter. 2006. The evolution of aging and age-related physical decline in mice selectively bred for high voluntary exercise. *Evolution* 60: 1494–1508.

Cespedes, A. M., and S. P. Lailvaux. 2015. An individual-based simulation approach to the evolution of locomotor performance. *Integrative and Comparative Biology* 55: 1176–1187.

Cespedes, A., C. M. Penz, and P. DeVries. 2014. Cruising the rain forest floor: butterfly wing shape evolution and gliding in ground effect. *Journal of Animal Ecology* 84: 808–816.

Chatfield, M. W. H., et al. 2013. Fitness consequences of infection by *Batrachochytrium dendrobatidis* in northern leopard frogs (*Lithobates pipiens*). *EcoHealth* 10: 90–98.

Dawson, T. J., and C. R. Taylor. 1973. Energetic cost of locomotion in kangaroos. *Nature* 246: 313–314.

Garland, T. 1983. Scaling the ecological cost of transport to body mass in terrestrial animals. *American Naturalist* 121: 571–587.

Hämäläinen, A., M. Dammhahn, F. Aujard, and C. Kraus. 2015. Losing grip: senescent decline in physical strength in a small-bodied primate in captivity and in the wild. *Experimental Gerontology* 61: 54–61.

Higham, T. E., and D. J. Irschick. 2013. Springs, steroids, and slingshots: the roles of enhancers and constraints in animal movement. *Journal of Comparative Physiology B: Biochemical, Systemic, and Environmental Physiology* 183: 583–595.

Hubel, T. Y., et al. 2016. Energy cost and return for hunting in African wild dogs and cheetahs. *Nature Communications* 7: doi 10.1038/ncomms11034.

Hunt, J., et al. 2004. High-quality male field crickets invest heavily in sexual display but die young. *Nature* 432: 1024–1027.

Husak, J. F. 2006. Does speed help you survive? A test with collared lizards of different ages. *Functional Ecology* 20: 174–179.

Husak, J. F., H. A. Ferguson, and M. B. Lovern. 2016. Trade-offs among locomotor performance, reproduction and immunity in lizards. *Functional Ecology* 30: 1665–1674.

Husak, J. F., A. R. Keith, and B. N. Wittry. 2015. Making Olympic lizards: the effects of specialised exercise training on performance. *Journal of Experimental Biology* 218: 899–906.

Killen, S. S., D. P. Croft, K. Salin, and S. K. Darden. 2016. Male sexually coercive behaviour drives increased swimming efficiency in female guppies. *Functional Ecology* 30: 576–583.

Kogure, Y., et al. 2016. European shags optimize their flight behaviour according to wind conditions. *Journal of Experimental Biology* 219: 311–318.

Lailvaux, S. P., and J. F. Husak. 2014. The life-history of whole-organism performance. *Quarterly Review of Biology* 89: 285–318.

Lailvaux, S. P., R. L. Gilbert, and J. R. Edwards. 2012. A performance-based cost to honest signaling in male green anole lizards (*Anolis carolinensis*). *Proceedings of the Royal Society of London B: Biological Sciences* 279: 2841–2848.

Lailvaux, S. P., F. Zajitschek, J. Dessman, and R. Brooks. 2011. Differential aging of bite and jump performance in virgin and mated *Teleogryllus commodus* crickets. *Evolution* 65: 3138–3147.

Lane, S. J., W. A. Frankino, M. M. Elekonich, and S. P. Roberts. 2014. The effects of age and lifetime flight behavior on flight capacity in *Drosophila melanogaster*. *Journal of Experimental Biology* 217: 1437–1443.

Magurran, A. E. 2005. *Evolutionary Ecology: The Trinidadian Guppy*. Oxford: Oxford University Press.

Marden, J. H. 1987. Maximum lift production during takeoff in flying animals. *Journal of Experimental Biology* 130: 235–258.

Murphy, K., P. Travers, and M. Walport. 2008. *Immunobiology*, 7th ed. New York: Garland.

Payne, N. L., et al. 2016. Great hammerhead sharks swim on their side to reduce transport costs. *Nature Communications* 7: 12289.

Pinshow, B., M. A. Fedak, and K. Schmidt-Nielsen. 1977. Terrestrial locomotion in penguins: it costs more to waddle. *Science* 195: 592–594.

Portugal, S. J., et al. 2014. Upwash exploitation and downwash avoidance by flap phasing in ibis formation flight. *Nature* 505: 399–402.

Reaney, L. T., and R. J. Knell. 2015. Building a beetle: how larval environment leads to adult performance in a horned beetle. *PLoS One* 10: e0134399.

Reznick, D. N., et al. 2004. Effect of extrinsic mortality on the evolution of senescence in guppies. *Nature* 431: 1095–1099.

Roberts, T. J., R. L. Marsh, P. G. Weyand, and C. R. Taylor. 1997. Muscular force in running turkeys: the economy of minimizing work. *Science* 275: 1113–1115.

Royle, N. J., J. Lindström, and N. B. Metcalfe. 2006. Effect of growth compensation on subsequent physical fitness in green swordtails *Xiphophorus helleri*. *Biology Letters* 2: 39–42.

Royle, N. J., N. B. Metcalfe, and J. Lindström. 2006. Sexual selection, growth compensation and fast-start swimming performance in green swordtails, *Xiphophorus helleri*. *Functional Ecology* 20: 662–669.

Rusli, M. U., D. T. Booth, and J. Joseph. 2016. Synchronous activity lowers the energetic cost of nest escape for sea turtle hatchlings. *Journal of Experimental Biology* 219: 1505–1513.

Spencer, R. J., M. B. Thompson, and P. Banks. 2001. Hatch or wait? A dilemma in reptilian incubation. *Oikos* 93: 401–406.

Ward, P. I., and M. M. Enders. 1985. Conflict and cooperation in the group feeding of the social spider *Stegodyphus mimosarum*. *Behaviour* 94: 167–182.

Weihs, D. 2002. Dynamics of dolphin porpoising revisited. *Integrative and Comparative Biology* 42: 1071–1078.

Williams, T. M., et al. 1992. Travel at low energetic cost by swimming and wave-riding bottlenose dolphins. *Nature* 355: 821–823.

———. 2014. Instantaneous energetics of puma kills reveal advantage of felid sneak attacks. *Science* 346: 81–85.

Wilson, R. P., B. Culik, D. Adelung, N. R. Coria, and H. J. Spairani. 1991. To slide or stride: when should Adélie penguins (*Pygoscelis adeliae*) toboggan? *Canadian Journal of Zoology* 69: 221–225.

Wyneken, J., and M. Salmon. 1992. Frenzy and postfrenzy swimming activity in loggerhead, green, and leatherback hatchling sea turtles. *Copeia* 1992: 478–484.

Zamora-Camacho, F. J., S. Reguera, M. V. Rubiño-Hispán, and G. Moreno-Rueda. 2015. Eliciting an immune response reduces sprint speed in a lizard. *Behavioral Ecology* 26: 115–120.

Chapter Nine: Nature and Nurture

Berwaerts, K., E. Matthysen, and H. Van Dyck. 2008. Take-off flight performance in the butterfly *Pararge aegeria* relative to sex and morphology: a quantitative genetic assessment. *Evolution* 62: 2525–2533.

Blows, M. W., et al. 2015. The phenome-wide distribution of genetic variance. *American Naturalist* 186: 15–30.

Bouchard, C., T. Rankinen, and J. A. Timmons. 2011. Genomics and genetics in the biology of adaptation to exercise. *Comprehensive Physiology* 1: 1603–1648.

Boyle, E. A., Y. I. Li, and J. K. Pritchard. 2017. An expanded view of complex traits: from polygenic to omnigenic. *Cell* 169: 1177–1186.

Bräu, L., S. Nikolovski, T. N. Palmer, and P. A. Fournier. 1999. Glycogen repletion following burst activity: a carbohydrate-sparing mechanism in animals adapted to arid environments? *Journal of Experimental Zoology* 284: 271–275.

Cullum, A. J. 1997. Comparisons of physiological performance in sexual and asexual whiptail lizards (genus *Cnemidophorus*): implications for the role of heterozygosity. *American Naturalist* 150: 24–47.

Denton, R. D., K. R. Greenwald, and H. L. Gibbs. 2017. Locomotor endurance predicts differences in realized dispersal between sympatric sexual and unisexual salamanders. *Functional Ecology* 31: 915–926.

Ginot, S., J. Claude, J. Perez, and F. Veyrunes. 2017. Sex reversal induces size and performance differences among females of the African pygmy mouse, *Mus minutoides*. *Journal of Experimental Biology* 220: 1947–1951.

Higgie, M., S. Chenoweth, and M. W. Blows. 2000. Natural selection and the reinforcement of mate recognition. *Science* 290: 519–521.

Kearney, M., R. Wahl, and K. Autumn. 2005. Increased capacity for sustained locomotion at low temperature in parthenogenetic geckos of hybrid origin. *Physiological and Biochemical Zoology* 78: 316–324.

Le Galliard J., J. Clobert, and R. Ferrière. 2004. Physical performance and Darwinian fitness in lizards. *Nature* 432: 502–505.

Marden, J. H., et al. 2013. Genetic variation in HIF signaling underlies quantitative variation in physiological and life-history traits within lowland butterfly populations. *Evolution* 67: 1105–1115.

McKenzie, E., et al. 2005. Recovery of muscle glycogen concentrations in sled dogs during prolonged exercise. *Medicine and Science in Sports and Exercise* 37: 1307–1312.

Mee, J. A., C. J. Brauner, and E. B. Taylor. 2011. Repeat swimming performance and its implications for inferring the relative fitness of asexual hybrid dace (Pisces: *Phoxinus*) and their sexually reproducing parental species. *Physiological and Biochemical Zoology* 84: 306–315.

Raichlen, D. A., and A. D. Gordon. 2011. Relationship between exercise capacity and brain size in mammals. *PLoS One* 6: e20601.

Rhodes, J. S., S. C. Gammie, and T. Garland. 2005. Neurobiology of mice selected for high voluntary wheel-running activity. *Integrative and Comparative Biology* 45: 438–455.

Saglam, I. K., D. A. Roff, and D. J. Fairbairn. 2008. Male sand crickets trade-off flight capability for reproductive potential. *Journal of Evolutionary Biology* 21: 997–1004.

Sharman, P., and A. J. Wilson. 2015. Racehorses are getting faster. *Biology Letters* 11: 20150310.

Sorci, G., J. G. Swallow, T. Garland, and J. Clobert. 1995. Quantitative genetics of locomotor speed and endurance in the lizard *Lacerta vivipara*. *Physiological Zoology*
68: 698–720.

Storz, J. F., J. T. Bridgham, S. A. Kelly, and T. Garland. 2015. Genetic approaches in
comparative and evolutionary physiology. *American Journal of Physiology: Regulatory, Integrative and Comparative Physiology* 309: R197–R214.

Chapter Ten: Mice and Men

Adusumilli, P. S., et al. 2004. Left-handed surgeons: are they left out? *Current Surgery*
61: 587–591.

Brooks, R., L. F. Bussière, M. D. Jennions, and J. Hunt. 2004. Sinister strategies succeed
at the cricket World Cup. *Proceedings of the Royal Society of London B: Biological
Sciences* 271: S64–S66.

Carrier, D. R. 1984. The energetic paradox of human running and hominid evolution.
Current Anthropology 24: 483–495.

Carrier, D. R., and M. H. Morgan. 2015. Protective buttressing of the hominin face.
Biological Reviews 90: 330–346.

Carrier, D. R., S. M. Deban, and J. Otterstrom. 2002. The face that sank the *Essex*:
potential function of the spermaceti organ in aggression. *Journal of Experimental
Biology* 205: 1755–1763.

Coren, S., and D. F. Halpern. 1991. Left-handedness: a marker for decreased survival
fitness. *Psychological Bulletin* 109: 90–106.

David, G. K., et al. 2012. Receivers limit the prevalence of deception in humans: evidence from diving behaviour in soccer players. *PLoS One* 6: e26017.

Faurie, C., and M. Raymond. 2005. Handedness, homicide and negative frequency-
dependent selection. *Proceedings of the Royal Society of London B: Biological Sciences* 272: 25–28.

Grouios, G., H. Tsorbatzoudis, K. Alexandris, and V. Barkoukis. 2000. Do left-handed
competitors have an innate superiority in sports? *Perceptual and Motor Skills* 90:
1273–1282.

Lailvaux, S. P., R. S. Wilson, and M. M. Kasumovic. 2014. Trait compensation and sex-
specific aging of performance in male and female professional basketball players.
Evolution 68: 1523–1532.

Liebenberg, L. 2008. The relevance of persistence hunting to human evolution. *Journal
of Human Evolution* 55: 1156–1159.

Lieberman, D. E., and D. M. Bramble. 2007. The evolution of marathon running: capabilities in humans. *Sports Medicine* 37: 288–290.

Lieberman, D. E., D. M. Bramble, D. A. Raichlen, and J. J. Shea. 2009. Brains, brawn,
and the evolution of human endurance running capabilities. In *The First Humans:*

Origin and Early Evolution of the Genus Homo, ed. F. E. Grine, J. G. Fleagle, and R. E. Leakey, 77–92. Berlin: Springer.

Little, A. C., et al. 2015. Human perception of fighting ability: facial cues predict winners and losers in mixed martial arts fights. *Behavioral Ecology* 26: 1470–1475.

Loffing, F., and N. Hagemann. 2015. Pushing through evolution? Incidence and fight records of left-oriented fighters in professional boxing history. *Laterality* 20: 270–286.

Morgan, M. H., and D. R. Carrier. 2013. Protective buttressing of the human fist and the evolution of hominin hands. *Journal of Experimental Biology* 216: 236–244.

Nickle, D. C., and L. M. Goncharoff. 2013. Human fist evolution: a critique. *Journal of Experimental Biology* 216: 2359–2360.

Palmer, A. R. 2004. Symmetry breaking and the evolution of development. *Science* 306: 828–833.

Perez, D. M., S. J. Heatwole, L. J. Morrell, and P. R. Y. Backwell. 2015. Handedness in fiddler crab fights. *Animal Behaviour* 110: 99–104.

Pollett, T. V., G. Stulp, and T. G. G. Groothuis. 2013. Born to win? Testing the fighting hypothesis in realistic fights: left-handedness in the Ultimate Fighting Championship. *Animal Behaviour* 86: 839–843.

Raymond, M., D. Pontier, A. B. Dufour, and A. P. Møller. 1996. Frequency-dependent maintenance of left handedness in humans. *Proceedings of the Royal Society of London B: Biological Sciences* 263: 1627–1633.

Schulz, R., and C. Curnow. 1988. Peak performance and age among superathletes: track and field, swimming, baseball, tennis, and golf. *Journal of Gerontology* 43: 113–120.

Van Damme, R., and R. S. Wilson. 2002. Athletic performance and the evolution of vertebrate locomotor capacity. In *Topics in Functional and Ecological Vertebrate Morphology*, ed. P. Aerts, K. D'Août, A. Herrel, and R. Van Damme, 257–292. Maastricht: Shaker.

Zilioli, S., et al. 2015. Face of a fighter: bizygomatic width as a cue of formidability. *Aggressive Behavior* 41: 322–330.

ACKNOWLEDGMENTS

Several colleagues, collaborators, and friends graciously took time out of their busy schedules to read sections of this book. Some even volunteered to read the entire thing, perhaps under the mistaken impression that I would not ask them to do so. I am grateful to Graham Alexander, Rob Brooks, David Carrier, Malcolm Gordon, Ray Huey, Jerry Husak, Duncan Irschick, Jonathan Losos, Ashadee Miller, Sheila Patek, Erik Postma, Michael Sadie, Owen Terreblanche, Glynnis Williams, and Robbie Wilson for their comments and criticisms on various chapters, and well-meaning lies as to the quality thereof. Many embarrassing errors were purged thanks to their vigilance, and without them this book would have been significantly worse. To my collaborators and students whose manuscripts went unread and emails went unanswered while I smashed my face into a keyboard, a thousand apologies; it's just as well you didn't need me after all. Candice Bywater interrupted a European vacation to analyze data(!) and provide me with answers to queries about her fiddler crab work, for which I am extremely grateful—I probably would not have been so accommodating! Alain Dejean, Tanya Detto, Jerry Husak, Michele Johnson, Rob Knell, Stan Lindstedt, Leeann Reaney, and Ewald Weibel kindly gave me permission to use their photographs and figures—thank you all.

I agonized for some time over how to best assign credit within the main text to the many excellent researchers whose work I drew upon. While I did not want the book to become a litany of names and institutions, I also thought it was important from time to time to highlight the people who actually did the research that I now get to pontificate about. I ultimately went about this as I tend to go about so many other things—wildly inconsistently. Thanks to all of my colleagues whose fascinating research contributed to this book; I hope I did it justice. Thanks also to my agent, the indomitable Russell Galen, who believed in the project from the start; Laura Jones Dooley, the best manuscript editor ever; and Jean Thomson Black, Michael Deneen, Margaret Otzel, and everyone else at Yale University Press for their hard work (and patience!).

Finally, I would like to especially thank my partner, Debbie Ramil, without whom everything would have been done much sooner.

Note: Page numbers in italics indicate illustrations.